Industrial Radiation Hazards
Deskbook

Industrial Radiation Hazards

DESKBOOK

Nicholas P. Cheremisinoff, Ph.D.
Paul N. Cheremisinoff, P.E.
Michael F. Teresinski, P.E.

TECHNOMIC
PUBLISHING CO., INC.
LANCASTER · BASEL

Published in the Western Hemisphere by
Technomic Publishing Company, Inc.
851 New Holland Avenue
Box 3535
Lancaster, Pennsylvania 17604 U.S.A.

Distributed in the Rest of the World by
Technomic Publishing AG

Printed in the United States of America
10 9 8 7 6 5 4 3 2

Main entry under title:
  Industrial Radiation Hazards Deskbook

A Technomic Publishing Company book
Bibliography: p.
Includes index p. 217

Library of Congress Card No. 87-50628
ISBN No. 87762-536-0

TABLE OF CONTENTS

PREFACE

Following the Soviet Nuclear reactor failure at Chernobyl, a renewed focus on nuclear reactor-safety arose in the U.S. and Western communities. In the U.S., work intensified on improving reactor safety and reliability following the U.S. reactor incident in 1979 at Three Mile Island, near Harrisburg, Pennsylvania. The U.S. since then has embarked into a new era in reactor technology built around liquid-metal reactors, more reliable pressured-water reactors, improved fuels, and more secure reprocessing and waste disposal. In many ways, the nuclear industry has set examples for safety-conscious operations and process designs that can conceptually be applied to non-nuclear, industrial operations.

There are a large variety of radiations utilized throughout the industrial and medical sectors, that fall into nonionizing and ionizing types. Engineers, technicians and other support personnel potentially exposed in these working environments in either continuous or infrequent bases must be educated in the potential dangers and safety practices required. This monograph was prepared as introductory reading to the potential hazards of industrial radiation practices. The volume discusses both nonionizing and ionizing forms of radiation in relation to typical industrial uses. The volume is intended to provide engineers, scientists and technicians, as well as the layman, an overview of radiation protection procedures, along with references for greater in-depth study for those who wish to dig deeper.

NICHOLAS P. CHEREMISINOFF
PAUL N. CHEREMISINOFF
MICHAEL TERESINSKI

PART I
Nonionizing Radiation

1/ APPLICATIONS *AND* CONTROL *OF* ELECTROMAGNETIC WAVES

CHARACTERISTICS OF ELECTROMAGNETIC WAVES

A wave can be described as an oscillating variation of some physical characteristic of matter or energy. A visible example of this is the action of waves on the surface of a body of water in which the individual molecules of water are physically in motion. Another example is sound waves, which are alternating compressions and rarefactions of the air or other substance through which they are traveling. Both of these type of waves involve a mechanical motion or displacement of particles of the substance, or medium, through which they are traveling. In contrast, electromagnetic waves consist of oscillating variations in electric and magnetic fields. Since these variations are of field only and do not involve motion of the particles of the medium through which the wave travels, electromagnetic waves may occur in a vacuum as well as in other media.

In order for wave motion to exist it must be generated by some type of oscillating source. The number of oscillations of this source in a given time span will equal the number of waves created in the same time span, one complete oscillation resulting in one complete wave. The number of waves (oscillations) per unit of time is called the frequency and the unit of time commonly used is the second. Once generated, the velocity at which the wave travels, or propagates, depends on the properties of the medium in which it travels. Since the waves are generated at a rate of some number of waves per second and these waves travel with a velocity measured in terms of distance per unit time, the physical dimension for the wave can be computed (i.e., the wavelength). A wavelength is the equivalent of the distance between succeeding crests of waves. The relationship between wavelength, frequency and velocity is

$$\lambda \text{ (wavelength)} = \frac{V \text{ (velocity)}}{f \text{ (waves per second)}} \qquad (1.1)$$

3

where velocity is in units of length per second, wavelength in the same units of length. Frequency is more commonly given in Hertz where one Hertz equals one wave, or cycle, per second.

There are an infinite number of electromagnetic waves, each having its own unique frequency and associated wavelength. Theoretically, they can exist starting with one Hertz and continuing with increasing frequencies. This range of frequencies is referred to as the electromagnetic spectrum. Certain frequency ranges of the spectrum are of practical interest either because of their uses or observed effects. The frequency ranges of interest are listed in Table 1.1. These bands are further subdivided into radar and microwave communication bands and are often identified with alphabetical notations, and they have no official status. Table 1.2 lists these designations.

Figure 1.1 provides a graphical description of wave motion using a circular coordinate system. If a body moving at constant speed around a circular path is viewed in the plane of the circle, it appears to oscillate along the diameter of the circle. If this oscillating motion along line CD is combined with uniform motion at right angles to this path as though the oscillating body were a pen making a trace on a strip of paper while the paper moves at a uniform velocity, the entire oscillation will be mapped out as shown in Figure 1.1(B). This resultant curve is the graphical representation of wave motion.

When the body in motion has completed one full revolution about the circle, its projection has moved from M to D, then to C, and finally back to M, and hence, it has completed an oscillation. The frequency is the number of oscillations per second completed. The amplitude of an oscillation is the maximum displacement and is denoted by either MD or MC. Note that it is equal to the radius (r) of the reference circle. The wavelength is the distance along the direction of propagation that it takes for one complete oscillation. In one oscillation the radius of the reference circle makes a complete revolution, and in the same time the wave advances uniformly a distance equal to its wavelength. Hence, the velocity of the wave is

$$v = f\lambda \tag{1.2}$$

where f = frequency in units of Hertz (cycles per second or oscillations per second).

Waves travel at characteristic speeds which depend on the type of wave and the particular medium in which the wave exists. For example, sound waves travel approximately 335.3 meters per second in air, but in water, the speed is about 1524 meters per second.

Electromagnetic waves travel much faster—approximately 3×10^8 meters per second in free space. Free space is a term often used in discussing electromagnetic waves and refers not only to a vacuum but also remoteness from any material substances which affect the waves. In other materials the velocity of

Table 1.1. Spectral bands of electromagnetic waves.

ELF	Extremely low frequencies	0–3 KHz
VLF	Very low frequencies	3–30 KHz
LF	Low frequencies	30–300 KHz
MF	Medium frequencies	300–3000 KHz
HF	High frequencies	3–30 MHz
VHF	Very high frequencies	30–300 MHz
UHF	Ultra high frequencies	300–3000 MHz
SHF	Super high frequencies	3–30 GHz
EHF	Extremely high frequencies	30–300 GHz

electromagnetic waves is less than free space but still high compared to observable speeds. In the earth's atmosphere, the velocity is only very slightly less than free space and this difference, for all practical purposes, is negligible.

The speed of electromagnetic propagation in a vacuum is the same as the speed of light (light being extremely high frequency electromagnetic waves). Its value is 299,793 kilometers per second (about 3×10^8 meters/second).

Since the oscillations of waves are periodic, or repetitious, they are characterized by a frequency which is the rate at which the waves are generated. Fre-

Table 1.2. Frequency band designations.

Band Designation	New Frequencies (GHz)		Old Frequencies (GHz)	
A	0.1 to	0.25		
B	0.25 to	0.50		
C	0.50 to	1.00	3.90 to	6.20
D	1.00 to	2.00		
E	2.00 to	3.00		
F	3.00 to	4.00		
G	4.00 to	6.00		
H	6.00 to	8.00		
I	8.00 to	10.00		
J	10.00 to	20.00		
K	20.00 to	40.00	10.90 to	36.00
K_a	Discontinued		26.00 to	40.00
L	40.00 to	60.00	0.340 to	1.55
M	60.00 to	100.00		
P	Discontinued		0.225 to	0.39
Q	Discontinued		36.00 to	46.00
S	Discontinued		1.55 to	5.20
V	Discontinued		46.00 to	56.00
W	Discontinued		56.00 to	100.00
X	Discontinued		5.20 to	10.90

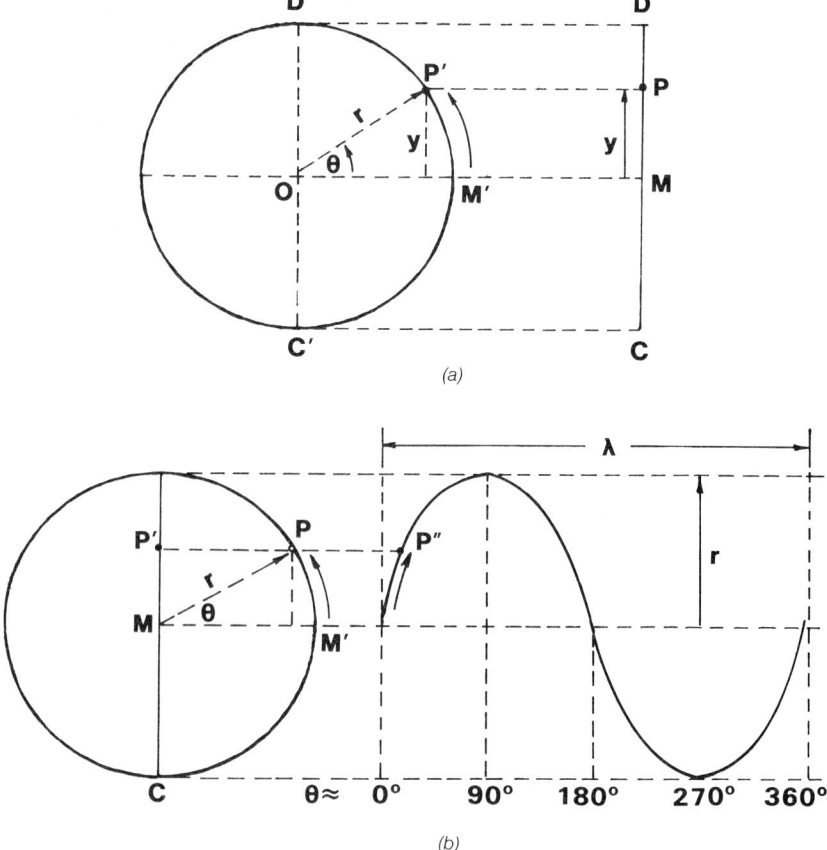

FIGURE 1.1. (a)—illustrates wave motion along circular coordinates; (b)—shows linear plot of circular motion.

quency is generally expressed in cycles per second, or Hertz, and is related to wavelength by the equation:

$$\lambda = \frac{v}{f} \qquad (1.3)$$

The wave velocity, v, has different values in different propagation media. The wave velocity depends on the media's permeability μ and permittivity ϵ:

$$v = \frac{1}{\sqrt{\xi\epsilon}} \qquad (1.4)$$

For free space the permeability and permittivity are respectively $\epsilon_0 = 1.26 \times 10^{-6}$ Henry/meter, and $\mu_0 = 8.85 \times 10^{-12}$ Farad/meter. Using these values in Equation (1.4) gives the free space velocity c:

$$c = \frac{1}{(1.26 \times 10^{-6})(8.85 \times 10^{-12})} = 3 \times 10^8 \text{ meters/second}$$

When the value in free space (c) is used, the resulting value of λ is the free space wavelength, λ_0, and Equation (1.1) gives

$$\lambda_0 = \frac{3 \times 10^8}{f} \tag{1.5}$$

where λ_0 is in meters. Note that as the frequency increases (decreased) the wavelength becomes shorter (longer) assuming that the propagation medium is the same.

FIELD INTENSITY AND POWER DENSITY RELATIONSHIPS

An electromagnetic wave can be broken down into an electric field and a magnetic field. Each component oscillates at the same frequency, travels with the same velocity, and has the same wavelength. Figure 1.2 shows the relationship between these two components.

E and H refer to the amplitude of the electric field and the magnetic field, respectively. The maxima and minima for the fields occur at the same time and place. This condition is known as being in phase. However, both fields are directed at right angles to each other and to the direction in which the wave is traveling or propagating, a relationship which is always maintained in free space. The fields are referred to as being mutually orthogonal.

Note that such a simple wave is formed only in the far field region. In the near field region E and H may be shifted in phase. The electric field is in phase with the electric charges and the magnetic field is in phase with the current in the radiator. An energy flow back and forth between these fields could occur. Therefore, the measurement of average power density is not valid in the near field region and E and H should be measured separately.

The wave shown in Figure 1.2 is referred to as being linearly polarized. That is, the electric field varies in amplitude in only one direction as it travels. In this drawing the wave is traveling in the z direction, the amplitude of the electric field varies only in the x direction, and the wave is said to be polarized in the x direction. By convention, polarization is described in terms of the electric field only and not the magnetic field. In actual space above the earth,

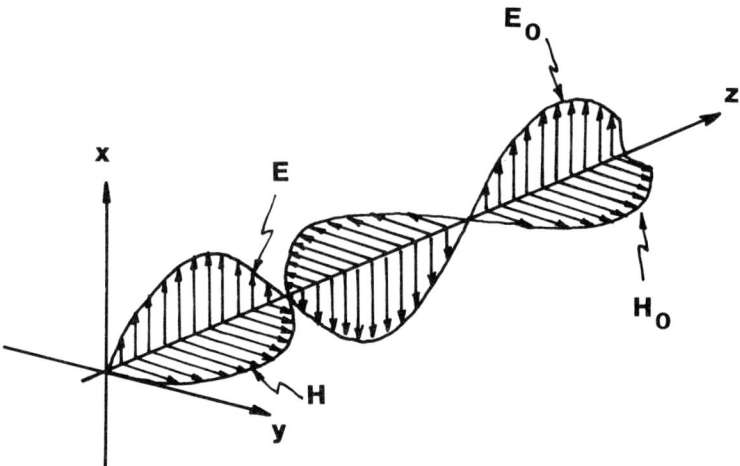

FIGURE 1.2. Schematic representation of relationship between electric and magnetic fields of an electromagnetic wave.

if the electric field is vertical, the wave is said to be vertically polarized; if the electric field is horizontal, the wave is said to be horizontally polarized.

Note that electromagnetic waves are not always linearly polarized. In circular polarization the electric field is rotating about the z axis as it travels so that the wave advances with a screw motion, making one full rotation about the axis for each wavelength it advances. The rotation may be clockwise or counterclockwise, corresponding to righthand circular and lefthand circular polarizations.

The polarization is random when there is no fixed polarization or pattern of polarization-variation that is repetitive along the direction of travel, an effect present in light waves emitted from an incandescent source such as the sun or a candle.

Electromagnetic waves are essentially energy flowing in the direction of propagation. The rate at which energy flows through a unit area in space (energy per unit time per unit of area) is the <u>power density</u> of the wave (expressed in watts/square meter, or milliwatts/square centimeter). If a source radiates power at a constant rate uniformly in all directions, the total power flowing through any spherical surface with its center at the source will be uniformly distributed over the surface and will equal the total power radiated. Such a source is ideal and is referred to as an <u>isotropic radiator</u>.

If a source radiates a fixed amount of power (P) isotropically, then the power density (W) at a distance r from the source will be the total radiated power divided by the area of a sphere with its center located at the source and having a radius r:

$$W = \frac{P}{A_s} = \frac{P}{4\pi r^2} \tag{1.6}$$

Similarly, the power density at a distance $2r$ is

$$W = \frac{P}{4\pi (2r)^2} = \frac{P}{16\pi r^2} \tag{1.7}$$

This value is less than the power density at distance r since $2r$ is larger than r. Consequently, the power density decreases as the distance from the source increases.

The power density of the field is proportional to the electric and magnetic intensities in a manner analogous to that of power in an electric circuit being related to the voltage and current (i.e., the product of the two). Since it is the average power density which is usually of interest, it is computed just as in AC circuits by multiplying the effective values of the electric (E) field and magnetic (H) field. The effective value is the maximum amplitude of the field multiplied by $1/\sqrt{2}$. Thus,

$$W = \frac{E_0}{\sqrt{2}} \times \frac{H_0}{\sqrt{2}} = \frac{E_0 H_0}{2} \tag{1.8}$$

where W is in watts/square meter when E_0 is expressed volts/meter and H_0 in amperes/meter.

Voltage and current in AC circuits are related through the resistance by Ohm's Law. Similarly, the electric and magnetic intensities are related through the characteristic wave impedance of free space, z_s, which is about 377 ohms (120π), a value which also holds for air. Therefore,

$$W = \frac{E^2}{377} = 377H^2 \text{ watts per meter}^2 \tag{1.9}$$

where E and H are the effective values ($E_0/\sqrt{2}$ and $H_0/\sqrt{2}$ in volts/meter and amperes/meter). It can also be shown that

$$H = \frac{E}{377} \tag{1.10}$$

for a wave in free space. Thus, by specifiying one field intensity, the other is automatically specified. Ordinarily, since the electric field is usually of most interest, the electric intensity is specified.

Power levels are reported in units of decibels (dB):

$$dB = 10 \, \text{Log} \, \frac{\text{Power}}{\text{Reference Power}} \qquad (1.11)$$

While the decibel system was primarily developed for use in acoustics, it is also applicable to the radio frequency portion of the spectrum, where it is widely used. In terms of power density the expression for decibels is

$$dB = 10 \, \text{Log} \left(\frac{W}{W_{ref}} \right) \qquad (1.12)$$

From the discussion of field strength and power density, it was determined that $W = E^2/377$, so Equation (1.12) can be restated as:

$$dB = 10 \, \text{Log} \, \frac{\left(\dfrac{E^2}{377} \right)}{\left(\dfrac{E_{ref}^2}{377} \right)} = 10 \, \text{Log} \left(\frac{E}{E_{ref}} \right)^2 \qquad (1.13)$$

From the law of logarithms, $\text{Log} \, N^2 = 2 \, \text{Log} \, N$, hence,

$$dB = 20 \, \text{Log} \left(\frac{E}{E_{ref}} \right) \qquad (1.14)$$

The decibel describes a ratio and because of this it is necessary to assume some reference value when using this unit. Any value may be selected as a reference value except zero. As long as this value is clearly specified, the expression of another value being X dB above or below the reference has a definite meaning. Since power or power density is one of the most important quantities in dealing with radio frequencies, particularly microwaves, the most used references are dBw (dB with respect to 1 watt), or dBm (dB with respect to 1 milliwatt). Figure 1.3 shows the relationship between dB and ratios. Although the absolute quantities used are the dBm and milliwatts per square centimeter (mW/cm²), the relationships will remain the same regardless of the quantities used.

Because the dB is a logarithmic scale, it doesn't add or subtract the way normal units do. For instance, if a 60 dBm signal is combined with another 60 dBm signal, the result is a 63 dBm signal. Sixty dBm corresponds to a power level of 1,000,000 mW and doubling this increases the level by a ratio of 2 to

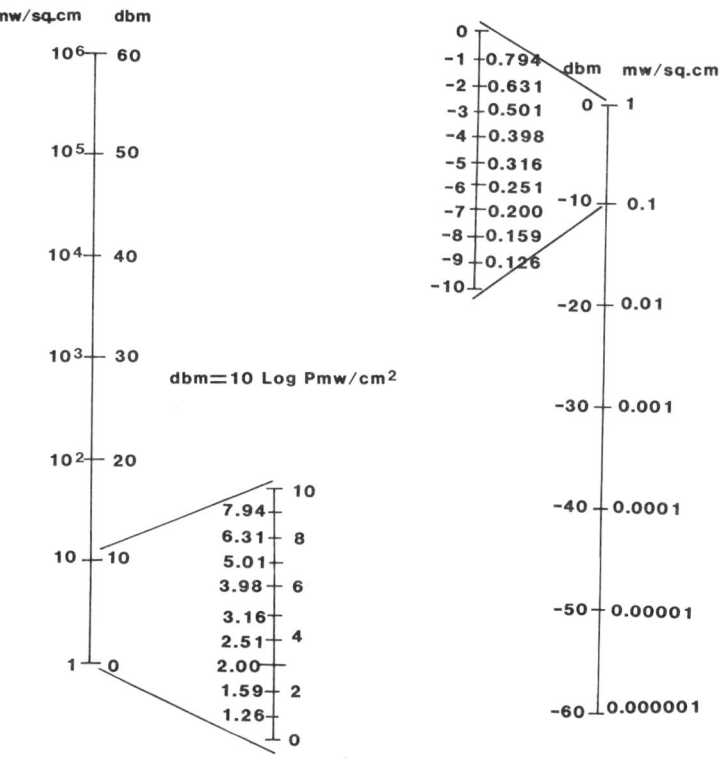

FIGURE 1.3. Shows relationship between power and decibels.

1. From the figure, it can be seen that a 2 to 1 ratio is a 3 dB increase, hence, the 63 dBm result.

WAVE PROPAGATION AND NATURE OF RADIATION

A half-wave antenna is one which is approximately a half-wave long at the operating frequency (e.g., at 276 megahertz, a half-wave antenna is approximately 5 meters long). When power is delivered to such an antenna, two fields are set up by the fluctuating energy: one the induction field, which is associated with the stored energy and the other the radiation field, which moves out into space at nearly the speed of light. At the antenna and a short distance from the antenna, the intensities of these fields are large and are proportional to the amount of power delivery to the antenna (beyond that short distance only the radiation field prevails). This radiation field is made up of an electric component and a magnetic component at right angles to each other in space and varying together in intensity.

The manner in which the radiation field is propagated away from the antenna is illustrated in Figure 1.4. The electric and magnetic components are represented here by separate sets of flux lines, which are at right angles to each other and to the radial direction of propagation. The diagram is illustrative of any plane containing the antenna, and applies for only a single instant of time. The magnetic flux lines are denoted as circular lines having the axis of the antenna as their axis, so that they appear as dots and crosses. The electric flux lines also are closed, or endless, lines in the present case, and consist of arcs of circles lying in planes containing the antenna and joined in the manner shown. These electric flux lines reverse direction at precisely the places where the magnetic flux lines reverse, and their density varies along radial direction in the way that the magnetic flux density varies. The electric flux density is everywhere proportional to the magnetic flux density.

Note that as time passes, the flux lines in Figure 1.4 expand radially with the velocity of light, and new flux lines are created at the antenna to replace those that travel outward. Thus, oscillating electric and magnetic fields are pro-

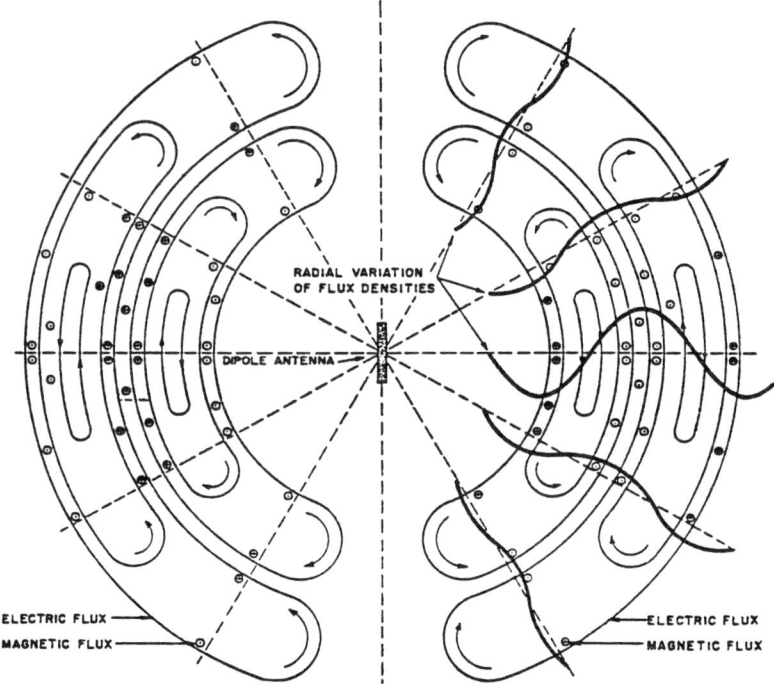

FIGURE 1.4. Illustrates electric and magnetic fields propagation from an antenna.

duced along the path of travel. The frequency of the oscillating fields is the same as the frequency of the antenna current, and the magnitudes of both fields vary continuously with this current. The variations in the magnitude of the electric component (called the E field) and those of the magnetic component (called the H field) are in time phase, so that at every point in space the time-varying magnetic field induces a difference in voltage, which is the electric field. The electric field also varies with time, and its variation is equivalent to a current, even though it is not associated with a movement of charge. This is referred to as the displacement current. It establishes a magnetic field in the same way that a conduction current does. Thus, the varying magnetic field produces a varying electric field, and the varying electric field, through its associated displacement current, sustains the varying magnetic field. Each field supports the other, and neither can be propagated by itself, without setting up the other.

The mechanism by which the flux lines of the radiation field become separated from the antenna and are radiated out into space can be understood by considering the movement of the charges which pass back and forth along the length of the antenna as a result of the driving current. At an instant of time when positive charges are distributed along one-half of the antenna, and negative charges along the other half, electric flux lines originate on the positive charges and terminate on the negative charges. These flux lines follow paths in space such as those indicated in Figure 1.5A. As time passes, the separated charges again come together, bringing the two ends of the flux lines together, as in Figure 1.5B. When the unlike charges meet, they seem to cancel each other, and the flux lines attached to them collapse and cease to exist. Thus, the two ends of the flux lines become joined, creating closed lines which are snapped free from the antenna and propagated outward, as shown in C and D of Figure 1.5.

We now direct attention to the phenomenon of wave propagation. There are two principal ways in which radio waves propagate from a transmitter to a receiver: by ground wave, or by free space radiation. Ground waves propagate by means of a conduction process, and are limited to those paths where both the transmitter and receiver are in actual or near contact with the earth. Ground wave propagation is utilized in the HF frequency region for short range communications.

Free space radiation is the most commonly employed mode of radio wave propagation. It is the only mode that is useable from aircraft, spacecraft or satellite, and is the preferred mode even in short range communications, where frequencies are high enough to achieve efficient antennas of reasonable size. Free space radiation is also the preferred mode at lower frequencies, where long range propagation is required, and large fixed antennas are possible.

Free space radiation can be divided into line of sight (straight line) and beyond line of sight (over-the-horizon) propagation. Line of sight is the principal

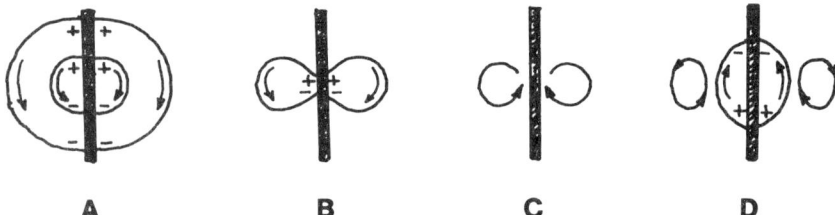

FIGURE 1.5. Characteristic closed electric lines at an antenna.

propagation mode for portable and vehicular communication sets and micro-wave communication networks. This mode is also used in most short range aircraft communication and navigation systems. All spacecraft and satellite communication networks are essentially line of sight propagation.

Free space radiation beyond line of sight is achieved by using the reflection, refraction and diffraction properties of the atmosphere at appropriate frequencies to bend the radio wave around the surface of the earth. Typical systems that employ beyond line of sight capability include: troposcatter and diffraction scatter microwave communication systems, and HF ionospheric communications. Ionospheric communication is probably the oldest known method of over-the-horizon radio and is commonly referred to as skywave communications.

The reflection of radio waves is like that of any other type of wave. When a beam of light falls on the surface of a mirror, nearly all of it is turned back or reflected. As with light waves, the efficiency with which reflection of radio waves occurs depends upon the material of the reflecting medium. Large, smooth, metal surfaces of good electrical conductivity (e.g., copper) are very efficient reflectors of radio waves. The surface of the earth is a fairly good reflector of radio waves, particularly for waves that are incident at small angles from the horizontal; and the ionosphere, even though it is not a surface such as a mirror, is also a fairly good reflector of radio waves.

Now consider a beam of light shining on a smooth surface of water. Some of the light will be reflected while the remaining portion will penetrate the water. The phenomenon by which light waves penetrate the water is referred to as refraction. The phenomenon can be readily observed by examining a glass of water into which an object is immersed. If viewed from an angle, the object appears broken or bent at the point where it enters the surface of the water. The reason for this is that light waves travel at a slower speed through water than through air. Thus, the direction of travel of the refracted light is different from that of the light beam incident on the surface of the water. A wave front is a surface of equal phase perpendicular to the direction of the travel of the wave. In the case of water waves the crests of adjacent circular expanding ripples would correspond to the wave fronts.

Refraction occurs only when the wave or beam of light approaches the new medium in oblique directions. If the entire wave front arrives at the new medium at the same moment, such as perpendicular to the surface, it is slowed up uniformly and no bending occurs. The amount any wave is refracted or bent as it passes from one medium to another is referred to as <u>refractive index</u>. This index depends on the relative densities of the two media. Refractive index is actually a ratio which compares the velocity of an electromagnetic wave through a perfect vacuum to its velocity through a denser medium (e.g., the earth's atmosphere).

Now consider a beam of light in an otherwise blacked-out room, shining on the edge of an opaque screen. In this experiment, it can be observed that the screen will not cast a perfectly outlined shadow. The edges of the shadow are not outlined sharply because the light rays are bent around the edge of the object and decrease the area of total shadow. The diffraction or bending of a light wave around the edge of a solid object is slight. The lower the frequency of the wave, or the longer the wavelength the greater the bending of the wave. Thus, radio waves are more readily diffracted than light waves, and sound waves more than radio waves. This phenomenon helps explain why radio waves of the proper frequency can be received on the far side of a hill or other natural obstruction, and why sound waves can be heard readily from around the corner of a large building. Diffraction is an important consideration in the propagation of radio waves at a distance, because the largest object to be contended with is the bulge of the earth itself, which prevents a direct passage of the wave from the transmitter to the receiver.

As electromagnetic energy propagates from a transmitter to a receiver, it will undergo a reduction in power level (referred to as <u>attenuation</u>. The major sources of attenuation are divergence of the radio beam, multipath interactions, and energy dissipation through absorption losses. This last source usually results in the conservation of electromagnetic energy into a heat rise in an absorption medium.

Power transmitted from an ideal point source propagates in all directions with equal magnitude. An example of this is a point of light filling space. As the distance increases between the beam source (transmitter) and measuring point (receiver), the amount of power arriving from the source is rapidly attenuating—the result of beam <u>spreading or divergence</u>. This beam divergence attenuation is related to the surface area of a sphere which has its center at the transmitter and its radius equal to the range from the transmitter to the receiver. The power available at a receiver from a given transmitter is reduced by a factor of 1/4 each time the range between the transmitter and receiver is doubled. This factor is the divergence attenuation factor referred to as the $1/R^2$ attenuation factor. Expressed in dB, a power reduction of .250 is

$$\text{Attenuation (dB)} = 10 \text{ Log}_{10} .250 = -6 \text{ dB}$$

When electromagnetic energy arrives at a receiver antenna from two or more different directions, both destructive and constructive interference effects will normally result. This is commonly referred to as a multipath effect, and when the interference is destructive, the result is multipath attenuation. A good example of such an interference is the "ghost" effect often seen on TV screens which occurs when power arriving directly from the transmitter is interfered with by power arriving from a reflecting body. Such reflection sources as other buildings, powerlines, bodies of water and aircraft are common. Multipath attenuation is a significant factor in many line of sight and troposcatter communication systems, and attenuation levels can approach 100% or total cancellation of the desired signal.

As noted earlier, whenever propagating electromagnetic energy passes through a material medium, it interacts with that medium (i.e., reflection, refraction, diffraction). Many material substances are significant absorbers of electromagnetic energy at selected frequencies, polarization and incident angles, etc. Absorption attenuation describes the power loss related to this absorption mechanism, water vapor and various gases and vapors in the atmosphere are significant absorbers of electromagnetic energy. Water in the ground and sea water, as well as mineral deposits and substances within the earth also attenuate significantly at higher frequencies. While absorption attenuation is not generally as predictable as other modes of attenuation, it is significant, and must be considered in radar propagation networks.

There are additional properties of electromagnetic waves that are of scientific interest. The intent of this chapter has been to briefly review principal characteristics and to orient the reader for the first part of the book. The references listed at the end of this chapter provide in-depth discussions of electromagnetic wave properties and general wave theory/phenomena.

REFERENCES

1 Cornillault, J. *Appl. Optics,* 11:265 (1972).

2 Illinger, K. H., ed. *Biological Effects of Nonionizing Radiation.* Washington, DC: American Chemical Soc. (1981).

3 Kerker, M. *The Scattering of Light and Other Electromagnetic Radiation.* New York:Academic Press (1969).

4 Uman, M. F. *Introduction to the Physics of Electronics.* Englewood Cliffs, NJ: Prentice-Hall Publishers, Inc. (1974).

5 Ziemer, R. E. and W. H. Tranter. *Principles of Communications.* Boston, MA: Houghton Mifflin Co. (1976).

2 /OCCUPATIONAL HAZARDS *OF* RADIO FREQUENCIES *AND* MICROWAVES

RADIO FREQUENCY DIATHERMY AND OTHER APPLICATIONS

Electromagnetic waves have been used for generating thermal energy as early as the turn of the century. The type of heating is based on specific frequencies as designated by the Federal Communications Commission for industrial, scientific and medical applications, known as ISM bands. The ISM bands are listed in Table 2.1. Below frequencies of 100 MHz, radio frequency heating has wide applicability in both industry and the medical field. The most commonly used frequencies are 13.56 MHz, 27.12 MHz and 40.68 MHz. A radio frequency heat generator has three basic components. The first is a rectifier unit where electric power (AC) is converted into direct current (DC) for the second component, the oscillator, which transforms the direct current into high frequency (called radio frequency or RF) to provide power for the third component and working circuit. In the working circuit the power is used to heat matter which is placed between the electrodes of a capacitor or in or near a coil of copper tubing.

The source of a radiofrequency power is very similar to a small broadcasting station. The electric field generated from the power source generates a rapid vibration of the molecules such as that in a dielectric material, thus producing thermal energy.

Note that the electric and magnetic fields are vector quantities and they have a polarization or direction. The orientation of these vectors with respect to a human who is being exposed, is an important parameter in determining the amount of energy being absorbed.

Since the wavelength below 100 MHz is quite long, workers engaged in a heat sealing process are generally within one wavelength distance from the source. The terminology of near and far field indicates the distance from the source in wavelengths. At 27 megahertz the one wavelength is about 10 meters. This is the near field region, beyond that distance is about where the far field of a radiofrequency sealer begins.

Table 2.1. ISM frequencies assigned by the FCC.

13.56 MHz	±	6.78	kHz
27.12 MHz	±	160	kHz
40.68 MHz	±	20	kHz
915 MHz	±	25	MHz
2,450 MHz	±	50	MHz
5,800 MHz	±	75	MHz
22,125 MHz	±	125	MHz

In the near field region, there are certain properties that are important from a safety standpoint. First, the ratio of electric to magnetic fields is not constant. It varies from point to point in space. The fields are nonuniform in space. Thus, for a human body exposed in the near field region, only certain parts of the body receive intense exposure.

In the far field region, the relationship between the electric and magnetic field is constant. Entire body exposures are typical for persons in the far field region because they are in a spatially uniform field.

The power density is only valid in the far field and not in the near field. However, since power densities are commonly used, they are really equivalent to far field power density. The true units are electric and magnetic fields and the ten milliwatt per square centimeter standard refers to about 200 volts per meter and 0.5 amps per meter.

There are two types of electromagnetic-radiation-emitting devices which operate at near the three bands of frequencies, induction heaters and dielectric heaters.

Induction heating refers to the generation of heat in a conducting material by means of magnetically induced currents. When an alternating current flows in a conductor, an alternating magnetic field is produced in the surrounding area. Similarly, if any conducting materials are placed in an alternating magnetic field, a current flow exists in that material. This current is such that the countermagnetic field generated by it will tend to cancel the existing field. In other words, the material tends to heat up as it resists the flow of induced high frequency current. This method of heating is employed when conducting material such as metal is to be heated. The metal is positioned either inside or near the induction coil and it becomes hot at the surface and then progressively toward the center. Induction heating is typically employed in metalworking plants in such operations as hardening gear teeth, cutting tools and bearing surfaces. It is also used for annealing, soldering and brazing.

Dielectric heating is used for nonconducting materials. Most dielectric materials absorb energy when placed in a high voltage AC field. These nonconductors do have some degree of conductivity, due mainly to molecular rotation (dielectric loss) and movement of free ions (conduction loss). Consequently, these materials are referred to as <u>Lossy Dielectrics</u>. The degree of lossiness for a given material is a function of the frequency of the electric field. The absorbed energy from the electric field heats the absorber.

Most dielectric heating is accomplished at frequencies of 10–100 MHz. The material to be heated is placed into the high frequency electric field, such as between two parallel plates of a capacitor. The efficiency of such a system is around 50–60%. Typical applications of dielectric heating include woodworking plants for bonding polywood, paper, laminating, and general gluing operations. The food industry uses it for sterilizing containers and killing bacteria in foods. Other manufacturing operations include molding plastics, thermosetting, curing vulcanizing rubber, setting twist in textile materials, and binding and printing.

INDUSTRIAL HAZARDS

The potential hazards in the operation of radio frequency equipment are electric shock, burns and radiation exposure to operating, testing and maintenance personnel. Some of the potential sources of injuries include improper and/or lack of grounding, defective or inoperative interlocks, controls and relays, inadequate shielding of RF radiation areas, unauthorized adjustments of controls, failure to shut off power and use lockout procedures before servicing, failure to discharge capacitors, and failure to short high-voltage lead to ground before working on equipment. Other hazards such as the effects of radio frequency radiation on biological matter are described later.

For diathermy, to safeguard against potential hazards in the use of RF/microwave diathermy units, one should avoid inadvertent exposure of the eyes of either personnel administering the treatment or the patient to the RF/microwave radiation.

Additional recommendations for further control of the units include:

- Calibrate the diathermy units in accordance with the Accreditation Manual for Hospitals, published by the Joint Commission of Accreditation of Hospitals.

- All applications of RF microwave radiation to personnel should be performed only at the request of the attending physician or physical therapist.

- The suggested percentage of power setting should be maintained.

- The prescribed duration and areas of exposure should not be exceeded and should be administered by qualified personnel only.

- All applications in the facial area should be avoided and other types of treatment substituted.

- The diathermy unit should be secured when not in use.

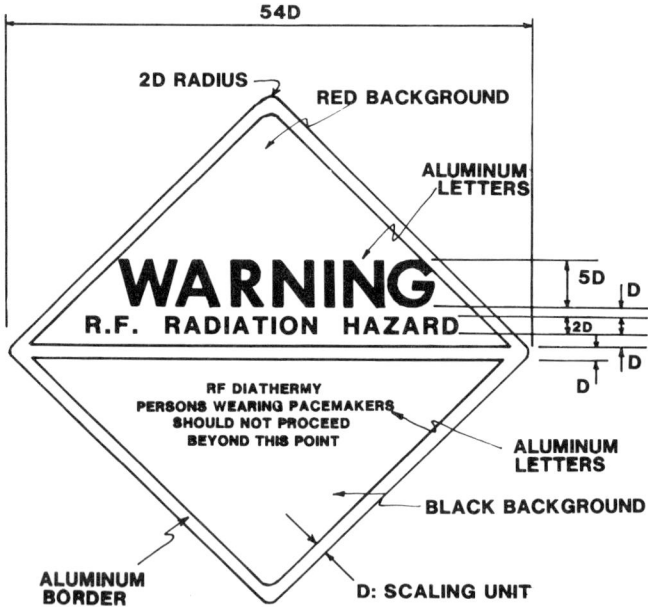

FIGURE 2.1. General RF radiation warning sign.

- The power setting control should be checked prior to operation to determine any looseness in the control knob position.
- The timing mechanism should be checked prior to each application to ascertain that the unit does cut off at the end of the preset time of exposure.
- The patient should be cautioned to avoid looking at the director during periods of application and not to place his face close to the director.

General diathermy operating procedure should be to:

- Make the patient ready and comfortable.
- Place the director in the vicinity of the area to be treated.
- Set all controls necessary for warm up and operation of the unit.
- Be certain persons in adjacent beds or areas will not be exposed.
- Be certain pacemaker patients are not in the vicinity of the unit.
- Apply RF power.
- Check settings and position of the director.
- Caution patient not to place face in field of radiated energy from the director.

A general RF radiation warning sign such as shown in Figure 2.1 should be positioned at the entrance to the therapy area when the unit is in operation.

NIOSH/OSHA GUIDELINES

The National Institute for Occupational Safety and Health (NIOSH) and the Occupational Safety and Health Administration (OSHA) have established guidelines outlining the potential health hazards to workers exposed to radiofrequency (RF) energy emitted from RF dielectric heaters. RF energy is non-ionizing electromagnetic radiation and should not be confused with X-rays and other ionizing radiation. RF energy, when absorbed in sufficient amounts by workers, may produce adverse thermal effects resulting from heating of deep body tissue which may include potentially damaging alterations in cells. Absorption of RF energy may also result in "nonthermal" effects on cells or tissue, which may occur without a measurable increase in tissue or body temperature. "Nonthermal" effects have been reported to occur at exposure levels lower than those that cause thermal effects.

OSHA notes that workers near RF sealers may be unaware of their exposure to RF emissions, because the RF energy from sealers and heaters can penetrate deeply into the body without activating the heat sensors located in the skin. A false sense of employee safety may exist; in many instances, worker exposures to RF energy may not have been properly assessed. This has been due, largely, to the complex problems of measurement and thus the misapplication of the instruments available for monitoring RF energy levels. Recently, monitoring instruments that facilitate accurate measurement of worker exposure have been developed.

RF sealers are used to heat, melt, or cure materials such as plastic, rubber, or glue. Specific uses include:

1. The manufacture of many plastic products such as toys, vinyl loose-leaf binders, rain apparel, waterproof containers, furniture slipcovers, and packaging materials.
2. Wood lamination and veneer processes, including glue setting.
3. Embossing and drying operations in the textile, paper, plastic and leather industries.
4. Curing of various materials including plasticized polyvinyl chloride, wood resins, polyurethane foam, concrete binder materials, rubber tires, and epoxy resins.

In the measurement of RF energies, the distances from the RF source at which the measurements are being made must be considered. NIOSH categorizes distances from the RF source as being either far field or near field. The far field includes all distances from the RF source greater than approximately ten times the wavelength. Wavelengths corresponding to frequencies used by RF sealers and heaters may range from about one meter to a few hundred meters. The frequency of 27 MHz, which is typical for many RF heaters, is associated with a nominal wavelength of about 11 meters. The value of the power density in the far field can be measured with a power density monitor, or can be calculated from measurement of the intensity of either the electric field or the magnetic field alone.

The near field comprises distances from the RF source less than about five wavelengths, which includes the immediate vicinity of the RF device where most worker exposures to RF energy occur. In the near field, electromagnetic waves have different characteristics than in the far field. Furthermore, in the occupational setting near a RF sealer or heater, the electromagnetic field generally is not uniform, and the energy field incident upon a worker is complex. A power density monitor, designed for use in the far field, is likely to give exceedingly inaccurate measurements in the near field. Further, in the near field, as opposed to the far field, there is no simple mathematic equivalency between values of power density and measurements of either electric or magnetic field strength.

When RF energy propagating through space intercepts an object, it may be reflected by the object, transmitted through the object, or absorbed. The extent to which RF energy is reflected, transmitted, and/or absorbed depends on the frequency of the RF energy, and on the shape, size, and dielectric properties of the object as well as its orientation relative to the incident RF energy. Human beings absorb RF energy at the frequencies used by most RF sealers and heaters. In workers who are not in contact with an electrical ground, the highest absorption rates for whole-body irradiation can occur at frequencies between 60 and 100 MHz with a peak at approximately 80 MHz. These frequencies of high absorption rates are very close to the frequencies used by most sealers and heaters. Hence, workers near RF sealers and heaters can absorb considerable amounts of the stray energy emitted from the RF machines. Effects of directly touching an electrical ground plane can lower, by as much as one-half, the frequency at which an irradiated body will maximally absorb energy. Contact of workers with an electrical ground plane can shift the frequency of maximum absorption rate to well within the frequency band of most RF sealers and heaters. This could increase the amount of energy absorbed by the worker and worsen the exposure condition. RF shielding material incorporated into the floor, walls, and ceiling of some RF workrooms could constitute such a ground plane.

NIOSH outlines the control of the emission of RF energy from RF sealers and heaters, stressing the use of properly designed shielding material. The shielding should be placed on or around the equipment so as to minimize occupational exposure due to emissions of stray RF energy. All shielding material should be properly grounded. Shielded conductors should be used for conveying RF current, and path impedance should be minimized by using good conductor materials. Many of these control features are available on RF sealers and heaters being marketed new, and some machines already in use can be retrofitted with some of these properties. NIOSH further recommends that the distance between the worker and the source of RF energy emission should be maximized. Examples of means to accomplish this include the use of automatic feeding devices, rotating tables, and remote materials handling. Also, the RF sealing and heating equipment should be electronically tuned to minimize the stray power emitted.

Whenever possible, equipment should be switched off when not being used. Maintenance and adjustment of the equipment should be performed only while the equipment is not in operation. After the performance of maintenance or repair, all machine parts, including cabinetry, should be reinstalled so that the equipment is intact and its configuration is unchanged.

Access to the vicinity of RF sealers and heaters where there may be stray RF energy should be limited as much as possible to the operator and necessary assistants, maintenance personnel, and industrial hygiene or safety personnel. Use of the RF equipment should be restricted to properly trained personnel.

Areas in which exposures to RF energy have been determined to be appreciable should be posted. Any signs should be of such size as to be recognizable and readable from a distance of three meters. All warning signs must be printed in English and in the predominant languages of non-English-reading workers, and should conform to the design recommended by OSHA.

Areas in which the RF energy is present at levels higher than the permissible exposure limit also should be posted. The warning signs should contain the following additional information: HAZARD – DO NOT ENTER. The sign must be readable from a distance of three meters. The perimeter of the restricted area should be clearly demarcated with signs visible to all personnel approaching the area.

A medical surveillance program should be developed. The program should include preplacement examination of all new employees and an initial examination of all present employees subject to occupational exposure to RF energy. In an effort to identify possible adverse effects associated with exposure to RF energy, annual examinations should be considered for workers who may be exposed to RF energy on a regular, long-term basis. Work histories should be included in all examinations.

Medical histories and physical examinations should give particular emphasis upon target organs potentially affected by RF energy including the eye (cataracts), the central nervous system, the blood (decreased leukocyte count), the immune defense system, and the reproductive system. Adverse reproductive effects may involve both men and women. For persons occupationally exposed to RF energy, medical records including health and work histories should be maintained throughout the period of employment and for an extended period after termination of employment.

Areas in the occupational environment where levels of RF energy have been determined to be appreciable should be surveyed at regular intervals. Immediately following a physical or electronic alteration of the equipment or an alteration in the process, a complete survey should also be performed. If measurements taken during a survey indicate that occupational exposure exceeds the permissible exposure limit, a second survey should be made on the next workday. If the limit is still exceeded, the use of RF equipment producing excessive values should be prohibited until appropriate controls have been instituted. The survey data sheets should contain all information pertaining to the survey, and should include the date and time of measurement, the type of monitoring equipment used, the employees' names, and the remedial actions taken, if any. These records should be maintained for an extended period of time.

References at the end of this chapter provide further information and guidelines.

REFERENCES

1 Conover, D. L., W. H. Parr, E. L. Sensintaffer and W. E. Murray, Jr. "Measurement of Electric and Magnetic Field Strengths from Industrial Radiofrequency (15–40.68 MHz) Power Sources," in *Biological Effects of Electromagnetic Waves: Selected Papers of the USNC/URSI Annual Meeting* (Boulder, CO, October 20–23, 1975). C. C. Johnson and M. L. Shore, eds. Department of Health, Education and Welfare, Public Health Service, Food and Drug Administration, Bureau of Radiological Health, DHEW Publication (FDA) No. 77–8011: 356–362 (1976).

2 Conover, D. L. "RF (10–40 MHz) Personnel Exposure," *Industrial Hygiene Problems*. American Industrial Hygiene Association Conference, New Orleans, LA, May 22–27, 1977 (1977).

3 Demokidova, N. K. "The Effects of Radiowaves on the Growth of Animals," in *Biological Effects of Radiofrequency Electromagnetic Fields*. Z. V. Gordon. ed. Arlington, VA:U.S. Joint Publications Research Service No. 63321, pp. 237–242 (1974).

4 Durney, C. H., C. C. Johnson, P. W. Barber, H. Massoudi, M. F. Iskarder and J. C. Mitchell. *Radiofrequency Radiation Dosimetry Handbook, 2nd edition*. SAM–TR–78–22. Brooks Air Force Base, TX, Department of the Air Force, Air Force Systems Command, Aerospace Medical Division, School of Aerospace Medicine (1978).

5 Gandhi, O. P. "Conditions of Strongest Electromagnetic Power Deposition in Man and Animals." *IEEE Trans. Microwave Theory Tech.*, 23:1021–1029 (1975).

6 Joint NIOSH/OSHA Current Intelligence Bulletin No. 33, "Radiofrequency Sealers and Heaters: Potential Health Hazards and Their Prevention," Washington, DC (Dec. 4, 1979).

7 Lobanova, E. A. and A. V. Goncharova. "Investigation of Conditioned-Reflex Activity in Animals (Albino Rats) Subjected to the Effect of Ultrashort and Short Radio-Waves," *Gig. Tr. Prof. Zabol.*, 15(1):29–33 (1971).

8 Serdiuk, A. M. "Biological Effect of Low-Intensity Ultrahigh Frequency Fields," *Vrach. Delo,* 11:108–111 (1969).

9 Volkova, A. P. and P. O. Fukalova. "Changes in Certain Protective Reactions of an Organism Under the Influence of SW in Experimental and Industrial Conditions." in *Biological Effects of Radiofrequency Electromagnetic Fields*. Z. V. Gordon, ed. Arlington VA, U.S. Joint Publications Research Service No. 63321, 168–174 (1974).

10 U.S. Department of Labor, Occupational Safety and Health Administration: OSHA Safety and Health Standards, 29 CFR 1910.97. OSHA 2206, Revised, Washington, DC (Nov. 7, 1978).

3/BIOLOGICAL *AND* OTHER EFFECTS *OF* ELECTROMAGNETIC RADIATIONS

POTENTIAL HAZARDS OF MICROWAVES

Microwave wavelengths typically range from 10 meters to about 1 millimeter, with the respective frequencies varying from 30 MHz to 300 GHz. An OSHA standard defines the microwave frequency range as 10 MHz to 100 GHz. In the area of industrial, scientific and medical applications, the Federal Communications Commission has categorized the frequencies for industrial, scientific and medical (ISM) uses as listed in Table 3.1.

As noted earlier, the far field region is one in which the distance between the source and absorber is greater than $2D^2/\lambda$ or $D^2/2\lambda$ where D is the largest distance between point in source, such as the diameter of the antennas reflector, and λ is the wavelength of RF/MW radiator. In this region the power density is expressed in milliwatts/square centimeter and the inverse square law is generally valid. In the near field region where the distance between the source and absorber is less than $2D^2/\lambda$ or $D^2/2\lambda$, the size and geometry of the source are important and the strengths of the incident radiation should be measured independently relative to the electric field or magnetic field. The units of measurements for electric and magnetic fields are volts/meter (v/m) and amperes/meter (A/M), respectively.

Where the power absorbed by biological materials principally depends on the field strengths of external RF/MW fields, the latter quantity could also be used as a measure of the biological dose. The amount of RF/MW radiation absorbed by biological matter depends on several factors, including the frequency, the material's degree of hydration, and its dielectric constant (Ke), magnetic permeability (μ), and electrical conductivity (σ). The absorber's physical dimensions, its geometry, and its angle with respect to the incident radiation are also important parameters. The combination of curved surfaces and high dielectric constant can sharply focus RF/MW radiation.

The biological effects of RF/MW radiation generally can be divided into

27

Table 3.1. ISM frequencies assigned by the FCC.

13.56	MHz ±	6.78	kHz
27.12	MHz ±	160	kHz
40.68	MHz ±	20	kHz
915	MHz ±	25	MHz
2,450	MHz ±	50	MHz
5,800	MHz ±	75	MHz
22,125	MHz ±	125	MHz

thermal and nonthermal radiation. The thermal effects resulting from the heating of body tissues are similar to those produced by conventional thermal methods. The nonthermal effects involve small changes and they appear to result from a direct interaction between the electric field and biological substance.

Living tissue is comprised of fixed charged molecules, which have a dipole moment. These molecules form the structural part of tissues and organs. There are also movable charged particles, such as ions of the electrolytes and macromolecules that are dissolved in all biological fluids. When exposed to microwave radiation, these charged particles experience electric forces. These induced forces lead to flow in the case of the movable charged particles and consequently to Joule heating of the biological material. The induced electrical forces on the immobile molecules may also lead to heat production due to molecular vibration or rotation. Furthermore, the electric field-induced forces may change the spatial distribution of polar molecules from a random orientation to an orientation aligned with the electric field. In all cases of microwave irradiation of living organisms, both of these induced force effects occur simultaneously. These additional molecular motions result in greater kinetic energy of the molecules which is observed physically as a rise in temperature. When biological effects are due mainly to heating, the phenomenon is referred to as thermal effects. The current exposure standards in the Western hemisphere are based on the assumption that this thermal interaction with biological systems is the only important type of interaction.

For frequencies between 10 MHz and 10 GHz, the human organs that are most likely to be damaged by microwave radiation are the testicles and the eyes. The damage to testicles is due to the degeneration of the epithelial lining of the seminiferous tubules followed by a rapid reduction of maturing spermatocytes. The injury to the eyes materializes as the formation of cataracts. At frequencies in excess of 10 GHz, surface heating results.

Various medical reports from the Eastern block countries report that long-term exposures to microwave radiation produce the following symptoms: dullness, headache, partial loss of memory, irritability, emotional instability, insomnia, diminished intellectual capability and high rate of heart beat. Other

Table 3.2. Subjective effects on persons working in RF fields.

Headaches	Feelings of fear
Eyestrain	Nervous tension
Fatigue	Mental depression
Dizziness	Memory impairment
Disturbed sleep at night	Pulling sensation in the scalp and brow
Sleepiness in daytime	Loss of hair
Moodiness	Pain in muscles and heart region
Irritability	Breathing difficulties
Unsociability	Increased perspiration of extremities
Hypochondriac reactions	Difficulty with sex life

occupational health studies suggest that the effects increase with total exposure time, even if exposure is intermittent. That is, despite therapeutic treatments and withdrawal from working with microwave sources, upon returning to previous working conditions, the symptoms increased in severity. These symptoms can take the forms of autonomic vascular disturbances, cerebral and coronary insufficiency, ischemic heart disease and hypertension. For women working with microwaves, the biological effects are changes in menstrual patterns, retarded fetal development and an increasing incidence of miscarriages.

Another important biological effect related to temperature rises induced by

Table 3.3. Occurrence of symptoms in humans exposed occupationally to high electromagnetic fields (750 KHz*200 MHz).*

	Length of Employment			
	1–6 Years (Average 4.3) (73 Persons)		7–16 Years (Average 9.6) (73 Persons)	
	Percent of cases	Number of cases	Percent of cases	Number of cases
Headache	20.5	15	32.9	24
Disturbance of sleep	13.7	10	23.3	17
Fatigue	12.3	9	17.8	13
General weakness	7.0	5	12.3	9
Disturbance of memory	5.5	4	8.2	6
Lowering of sexual potency	5.5	4	8.2	6
Drop in body weight	2.7	2	12.3	9
Disturbance of equilibration	5.5	4	11.0	8
Neurological symptoms	0.0	0	15.1	11
Changes in ECG	17.8	13	28.8	21

*After data by Minecki (1964) as quoted by McRee (1972).

Table 3.4. Clinical manifestations of chronic occupational exposure of 525 workers to microwave radiation.*

Symptomatology

Bradycardia
Disruption of the endocrine-humoral process
Hypotension
Intensification of the activity of thyroid gland
Exhausting influences on the central nervous system
Decrease in sensitivity to smell
Increase in histamine content of the blood

Subjective Complaints

Increased fatigability
Periodic or constant headaches
Extreme irritability
Sleepiness during work

*After data by Letavet and Gordon (1960).

microwave radiation in biological media is cataract formation. The effects of microwave radiation have been applied to test animals, resulting in the production of cataracts in rabbits. The medical literature reports that microwaves may have a cumulative effect on the lens of the eyes if there is an insufficient time interval between doses for the eyes' repair mechanism to act.

The energy for nonthermal effects is generally considered to be the energy density at less that of 1 mW/cm². This type of radiation is referred to as microthermal heating. It is very localized and the temperature rise cannot be detected by conventional temperature measurement instruments. Although the medical literature is not explicit, there is evidence to suggest that microwaves interact directly with the central nervous system without detectable heating.

Tables 3.2 through 3.4 list some of the effects on the brain and central nervous system of persons exposed to radio frequency and microwave radiation.

ENVIRONMENTAL IMPACT ASSESSMENT

Until about the mid-1960s, industry limited the exposure of personnel to microwave radiation to power density levels no greater than 10 mW/cm. The maximum permissible exposure level was the same regardless of the exposure duration. The increasing power output levels in newer microwave systems as well as military installations employing radar, had combined to make this limit difficult to comply with. Modern criteria equate higher exposure levels with time of exposure.

The American National Standards Institute (ANSI) has adopted new standards which relate the <u>level</u> of exposure to the <u>time</u> duration of such exposure. The ANSI Standard, C95.1–1974, recommends a CW radiation protection guide of 10 mW/cm for incident electromagnetic energy of frequencies from 10 MHz to 100 GHz. The equivalent free space electric and magnetic field strengths are approximately 200 v/m rms and 0.5 A/m rms, respectively. For modulated fields, the power density is averaged over any 0.1 hour period. Microwave radiation can be classified into continuous (cw) and pulse mode. The pulse mode, such as radar, is considered to be more important due to higher power intensity.

In some instances it may be necessary to determine the radiated power density distribution versus range through a theoretical analysis of system performance parameters. This method of establishing hazard criteria is based on a worst-case scenario evaluation. Some system parameters such as frequency, azimuth and elevation coverage, physical antenna size, and absolute transmitter power levels, can be accurately measured and inputed to such an analysis. There are, however, a number of characteristics which are important to an exact radiation analysis but that do not yield to exact measurement. This is particularly true for reflector antenna systems, where transmission line characteristics, antenna gain and efficiency, half-power beam widths, side lobes, blockage, etc., are initially specified and then subject to change in both positive and negative directions with age, use, type of maintenance, care in handling, etc.

In general, there are two separate zones wherein dissimilar behavior of radiated energy is experienced (i.e., <u>near field</u> and <u>far field</u> zones). In the near field zone the energy is collimated in a beam approximately the same size as the antenna aperture, which oscillates sinusoidally in amplitude, with range. The locus of maxima basically follows a constant level throughout this zone. In the far field zone, the power density level decreases at a rate proportional to the inverse square of the range, i.e., I/R^2. There is no fine boundary at which a sharp change in behavior takes place, but there is a transition region referred to as the <u>intermediate field</u>. There are also different definitions of these zones. Depending upon the particular definition, the distance from the antenna to where the far field begins varies from $2D^2/\lambda$ to $D^2/2\lambda$ (where D = antenna diameter, and λ = wavelength of the radiated energy).

The U.S. Army, Environmental Hygiene Agency, recommends a semi-empirical approach to performing a hazard analysis. Figure 3.1 defines zones of radiation defined in this analytic approach. For circular antennas, the following formulas can be used in computing the field power densities:

$$R_1 = \frac{D^2}{5.66\lambda} \tag{3.1}$$

$$R_1 = \frac{D^2}{2.83\lambda} \text{ or } 2R_1 \tag{3.2}$$

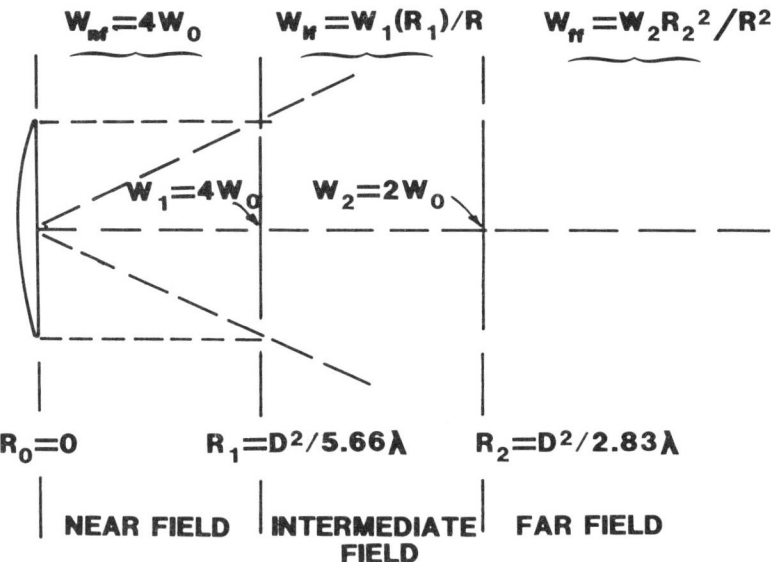

FIGURE 3.1. Zones of radiation.

where

R_1 = distance to end of near field
R_2 = distance to end of intermediate field
D = diameter of reflector

and

$$W_{nf} = 4W_0 \tag{3.3}$$

where

W_0 = average power density in near field
W_{nf} = maximum power density in near field

and

$$W_{if} = W_1(R_1/R) \tag{3.4}$$

$$W_1 = 4W_0 \tag{3.5}$$

$$W_2 = 2W \tag{3.6}$$

$$W_{ff} = W_2(R_2/R)^2. \tag{3.7}$$

where

W_{if} = power density of intermediate field at distance, R
W_1 = power density at end of near field
W_2 = power density at end of intermediate field
W_{ff} = power density in far field at distance, R

It is assumed that the maximum power density, W, exists throughout the near field (W_{nf}) to a distance, R_1, determined by $D^2/5.66\lambda$, where D is the diameter of the antenna reflector and λ is the wavelength of the transmission. The power density will then decrease approximately linearly to a distance, R_2, determined by $D^2/2.83\lambda$, at which point it will be W_2 or $2W_0$. From this point on it will decrease according to the inverse square law. Note that the units for D, R and λ must be consistent. Parameter R is always the distance from the antenna.

Although the actual power density will vary throughout the near field (W_{nf}), for personnel safety considerations the maximum that can exist ($4W_0$) is assumed to exist throughout the near field.

The maximum power density at the circular antenna reflector may be calculated as follows from Equation (3.3):

$$W = 4W_0$$

where

W = maximum power density at beam axis (mW/cm²)
W_0 = average power density across the antenna

and

$$W_0 = \frac{P}{A} \tag{3.8}$$

where P = average output power of transmitter (milliwatts). If peak power is given, P can be obtained by multiplying peak power by a duty factor. A = effective area of reflector (cm²).

As an alternate approach to computing the far field power density at a distance from the source, consider an isotropic radiator radiating in a homogeneous medium of infinite extent. Assume that this radiator radiates energy at a rate of P watts. At a distance r, the radiated power must be uniformly distributed over a spherical surface of $4\pi r^2$. The power density at any distance from the source is

$$W_{ff} = \frac{P}{4\pi r^2} \tag{3.9}$$

If the source is a transmitting antenna of gain G, radiating power at P watts, the power density at any point on the axis of the main lobe of the antenna is

$$W_{ff} = \frac{GP}{4\pi r^2} \qquad (3.10)$$

where r is the distance from the effective aperture of the transmitting antenna, along the axis of the main lobe and G, the gain, is

$$G = \frac{4\pi A}{\lambda^2} \qquad (3.11)$$

Combining Equations (3.11) and (3.10) yields the far field power density at a distance r along the major lobe:

$$W_{ff} = \frac{AP}{\pi^2 r^2} \qquad (3.12)$$

Equation (3.12) is known as the Friis free space transmission formula.

We now direct attention to the absorption of microwave radiation. Although photon energy in RF and microwave radiation is too low to produce photochemical reactions in biological matter, microwaves are absorbed by biological systems and will cause thermal damage to tissue.

Table 3.5 lists various parameters affecting human tissue exposed to an electric field. The first column lists selected frequencies between 27.12 MHz and 2,450 MHz. The frequencies of 27.12, 433, 915, and 2,450 MHz are significant since they are used for diathermy, microwave ovens, and various industrial heating operations. The frequency of 433 MHz is authorized only in Europe for these purposes. The second column tabulates the corresponding wavelength in air or the distance the wave will travel during the time of one cycle. The remaining columns pertain to the wave properties in two major classes of tissues. The first group consists of muscle, skin, or tissues of high water content, while the second group consists of fat, bone, and tissues of low water content. Other tissues containing intermediate amounts of water such as brain, lung, bone marrow, etc., will have properties that lie between the tabulated values for the two listed groups. The magnitude of the dielectric constants shown for each group in Table 3.5 has a major role in the reduction in wave velocity and wavelength as well as depth of penetration as the wave penetrates the tissue.

For a human body subjected to electromagnetic radiation, the body's conductivity (σ), dielectric constant (ke), and shape (radius of curvature) determine the amount of energy reflected from and transmitted through the surface.

Table 3.5. Electrical properties of human tissues.

Frequency (MHz)	Muscle					Fat			
	Wavelength in Air (cm)	Dielectric Constant Ke_m	Conductivity σ_m (mho/m)	Depth of Penetration (cm)	Wavelength in Tissue (cm)	Dielectric Constant Ke_f	Conductivity σ_f (millimho/m)	Depth of Penetration (cm)	Wavelength in Tissue (cm)
27.12	1006	113	0.612	14.3	68.1	20	10.9– 43.2	159	241
40.68	738	97.3	0.693	11.2	51.3	14.6	12.6– 52.8	118	187
100	300	71.7	0.889	6.66	27	7.45	19.1– 75.9	60.4	106
433	69.3	53	1.43	3.57	8.76	5.6	37.9–118	26.2	28.8
750	40	52	1.54	3.18	5.34	5.6	49.8–138	23	16.8
915	32.8	51	1.60	3.04	4.46	5.6	55.6–147	17.7	13.7
1500	20	49	1.77	2.42	2.81	5.6	70.8–171	13.9	3.41
2450	12.2	47	2.21	1.70	1.76	5.5	96.4–213	11.2	5.21

*Institute of Electrical and Electronics Engineers, Inc. "Therapeutic Applications of Electromagnetic Power," by A. W. Guy, J. F. Lehmann, and J. B. Stoneridge, *Proceeding IEEE*, Volume 62, January 1974, p. 61.

Of the power transmitted into the body's surface, σ, ke and curvature again determine the spatial distribution of absorbed energy. The absorbed energy is dissipated as heat in the tissue, and, in general, decreases exponentially with distance beneath the surface. In diathermy applications and in quantifying hazardous levels of radiation, it is important to quantify σ and ke, especially for the tissue near the surface of the body. Furthermore, a given surface area of the body may be preferentially heated by incident electromagnetic energy if that area presents σ and ke values appropriately different from the surrounding area.

By way of illustration, consider an electromagnetic wave normally incident in air upon a semi-infinite slab of biological matter. The incident energy transmitted, e, into the material is attenuated with distance x can be expressed as follows:

$$E(x) = E_{0e}^{-2\beta_x} \tag{3.13}$$

where $x = 0$ at the air-medium interface, and β is the attenuation constant of the medium.

At a distance $x = 1/\beta$ into the medium, Equation (3.13) reduces to

$$E(1/\beta) = E_{0e}^{-2} = 0.134E_0$$

This distance is referred to as the depth of penetration $= 1/\beta$. This is where the energy has diminished to 13.5% of the energy entering the medium. In other words, 86.5% of the original entering energy is absorbed within the penetrating depth.

The depth of penetration is expressed by

$$\delta = 1/\beta$$

where

$$\beta = 2\pi f \sqrt{\frac{\mu\epsilon}{2}} \left\{ \left[1 + \frac{\left(\frac{\sigma}{f_e}\right)^2}{4\pi^2} \right]^{1/2} - 1 \right\} \tag{3.14}$$

$$\sigma = \text{conductivity}$$

Through biological matter, the permeability $\mu = \mu_0 = 4\pi \times 10^{-7}$ Henrys/meter in free space. By substituting the value μ_c for μ, we obtain

$$\beta = 4.98 \times 10^{-3} f\sqrt{\epsilon} \left\{ \left[1 + 2.5 \times 10^{-2} \left(\frac{\sigma}{f\epsilon}\right)^2 \right]^{1/2} - 1 \right\}^{1/2} \tag{3.15}$$

where β has units of m^{-1}.

Biological matter is classified as lossy (i.e., it is essentially conductive). Since wavelengths will change as they enter a medium of different density, the internal wavelength in a lossy medium is

$$\lambda = \frac{\lambda_0}{\sqrt{K_0}} \left[1/2 + 1/2 \sqrt{1 + \left(\frac{\sigma}{\omega\epsilon} \right)^2} \right]^{-1/2} \tag{3.16}$$

or

$$\lambda = \frac{\lambda_0}{\sqrt{K_0}} \left[1/2 + 1/2 \sqrt{1 + \tan^2 \delta} \right]^{-1/2} \tag{3.17}$$

where $\tan \delta = \sigma/\omega\epsilon$.

These relationships indicate that the wavelength in a lossy medium is reduced.

To recap, microwave radiation is classified as nonionizing because a microwave photon is not strong enough to disrupt chemical bonds. However, microwave energy may be absorbed within the material and dissipated as heat. For a given material, if the ratio of $\sigma/(\omega\epsilon)$ is large, then most of the microwave energy incident from air would be reflected, and what little energy is absorbed would be deposited near the surface, due to the high attentuation rate within the material. If, however, $\sigma/(\omega\epsilon)$ is small, less energy is reflected from the surface, and the greater energy transmitted into the material would penetrate more deeply.

Microwave energy transmitted through curved surfaces (convex) of the body tend to be focused into a region beneath the surface, in the same way that an optical wave is focused by a lens. This accumulation of energy beneath curved surfaces, and the fact that microwaves are dissipated as heat, make the brain, the optic nerves, and the testes the most susceptible regions to over-exposure. Consequently, these are the areas which should be shielded if one must be within an area where a radiation hazard may exist. Since power generally decreases inversely with distance from the source squared, an obvious safety measure is to avoid close contact with high-power microwave sources. When this is not possible, the use of shielding or absorbing materials is required.

X-RAY GENERATION FROM MICROWAVE TUBES

Another potential hazard from microwaves is X-rays. X-rays are electromagnetic waves that are generated in a vacuum tube when high velocity electrons impinge on a metallic target. Several sources of this danger are X-ray machines which are extensively used by the medical and dental professions and in certain industrial applications. In addition, certain types of vacuum tubes used primarily in high power microwave and radar systems also generate

X-rays. In these systems, tubes, although not designed specifically to generate
X-radiation, radiate X-rays as an unwanted byproduct.

In microwave transmitting systems, certain electronic tubes, such as klystrons, magnetrons, traveling wave tubes, and high voltage thyratrons, contain
the basic physical parameters which allow them to behave as X-ray generators.
The key element is very high voltages required to operate the tubes utilized in
the generation of microwave energy. Microwave energy itself has never been
known to cause ionization of atoms, and, accordingly, cannot be a source of
such radiation.

There are, however, similarities between X-ray tubes and these electron
tubes. Figure 3.2 is a plot of relative X-ray intensity as a function of voltage.
Note that increasing the anode voltage from 20 kV to 40 kV increases the X-
ray intensity almost tenfold. Some klystrons employed in radar systems operate at potentials in excess of 250 kV.

Figure 3.3 illustrates tube components and similarities between X-ray tubes
and high voltage tubes. Figure 3.3(A) illustrates a typical X-ray tube employed
in medical and dental machines. Electrons are emitted from the cathode, are
accelerated by the high anode voltage, strike the target, and generate X-rays.

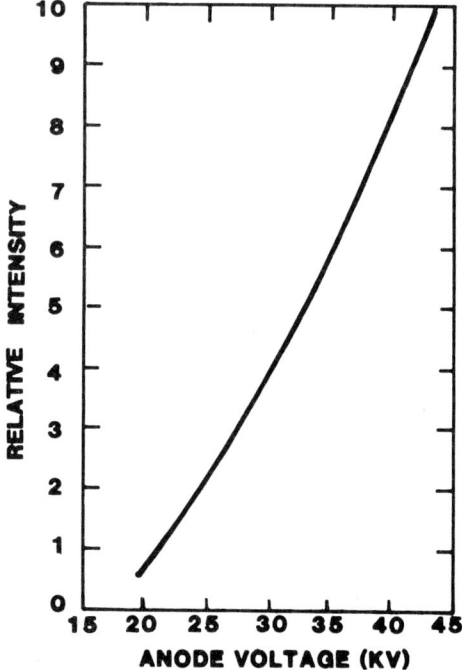

FIGURE 3.2. Plot of relative intensity of X-ray radiation vs. anode voltage.

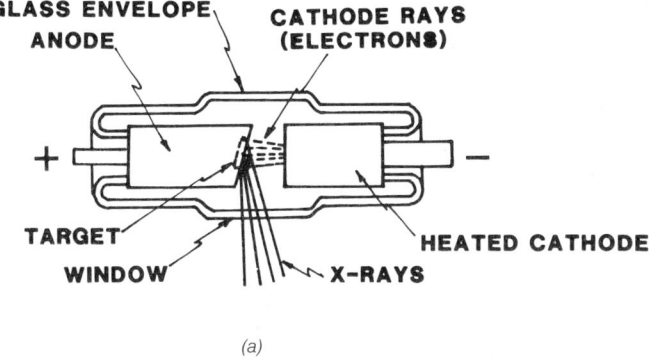

GLASS ENVELOPE
ANODE
CATHODE RAYS
(ELECTRONS)

+   −

TARGET
WINDOW
HEATED CATHODE
X−RAYS

(a)

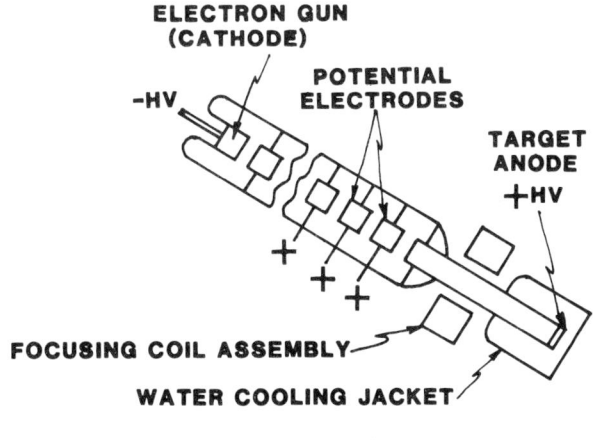

ELECTRON GUN
(CATHODE)
POTENTIAL
ELECTRODES
−HV
TARGET
ANODE
+HV

FOCUSING COIL ASSEMBLY
WATER COOLING JACKET

(b)

FIGURE 3.3. Illustrates (a) components of an X-ray tube; (b) high voltage multisection X-ray tube.

Figure 3.3(B) shows a higher voltage, multisection tube such as is used in X-ray therapy and certain industrial machines. These tubes typically operate in the megavolt range. This tube has several electrodes between the cathode and target anode. Each electrode is connected to a higher potential than the previous one. This arrangement accelerates the electrons from the cathode to the target and at the same time reduces the electric strain on the glass envelope which would be caused by electron bombardment.

A typical high voltage klystron tube is employed in a radar system. There are certain similarities between it and the high voltage multisection X-ray tube. In this tube, the electron beam generated by the electron gun assembly is focused by means of a magnetic field, is sent through the cavities, and is received by the collector assembly. Basically, the high velocity electrons bombard the collector assembly in much the same manner as those in the X-ray tube bombard the target element. The shaded area denotes the general distribution of X-radiation. The greatest intensity usually occurs near the collector assembly and output cavity.

HAZARDS FROM MICROWAVE COOKING OVENS

Microwave cooking ovens have become a popular home consumer product. Microwave cooking is based on intermolecular friction. When a substance is placed into the RF field generated by the magnetron, the molecules of the substance are interfered with by the varying electromagnetic field. This variation in the field causes increased movement of the molecules, resulting in the formation of heat. The amount of heating generated depends on the magnitude of the magnetron output power, the structure and the initial temperature of the material to be cooked. The magnitude of the required generator output power depends on the cooking requirements of the oven. An oven which is only used to reheat precooked food will require less power than one which must cook food completely. The size, configuration and density of the food plays a very important part in the amount of heat needed. A substance that is large in mass will require more heat than one which is thin or flat. Also, the initial temperature of the food to be heated is one of the most important factors influencing the energy requirements. Sufficient energy must be used to raise the temperature of the food from its initial level to the cooking temperature. Consequently, frozen food will either require more energy, or longer cooking time than food that is at room temperature.

In comparing the microwave oven to the conventional gas or electric oven, there are several unique differences. First, the process of preparing food by conventional cooking methods is very slow compared to cooking with microwaves. During conventional cooking methods, the surface of the food is heated; then by conduction, heat is carried to the center of the food. If the requirement is to speed up the cooking time, more heat must be added to the surface of the food. There is a limit to the amount of heat the surface can tolerate beyond which burning or scorching occurs. During the process of heating, the interior of the oven is also heated. In some cases, where insulation is inadequate, the exterior of the oven also will be heated. In microwave ovens, as soon as the microwave energy starts to penetrate, the conversion to thermal energy begins. As the energy progresses to its internal limit, the heating process continues. The normal penetration depth is generally found to be between 2–1/2 to 3 inches below the surface. However, this does not restrict the size of the item to be heated since conduction will carry heat to deeper areas. Microwave energy reaches the food directly from the source and from reflections by the walls and bottom of the oven, thus heating the food more uniformly. For this reason, it is important that containers holding the food be transparent to the microwave energy. Almost all glass, china, plastic, or paper containers are suitable, since they do not absorb or reflect microwave energy. Metal containers should not be used since they reflect the energy. During the process of cooking by microwaves, the interior walls do not absorb energy, and therefore, are not heated.

The Federal Communications Commission has assigned the frequencies 915 MHz and 2450 MHz for use in microwave cooking ovens. The frequency most commonly used is 2450 MHz. There are several parameters which affect the extent of damage to be expected from exposure to microwave radiation by these ovens. The first of these is frequency. The frequency range between 150 MHz and 1000 MHz is most hazardous since the energy is absorbed by deeper body tissue. Since the tissue below the skin layers has little temperature sensation, this range is even more hazardous because one would not immediately be aware of the damage being caused by the radiation. At 2450 MHz, a combination of the skin and deep tissue heating occurs, with the eye being the most susceptible to damage. Another important factor is the power density. Continual use of a unit eventually causes door seals to lose their shape, and consequently, leakage of radiation may increase with time. Power densities outside the oven may exceed 10 mW/cm^2. This increase may also be caused by the buildup of grease along the door seal and eventually failure of the door interlocks, which normally are adjusted so that unless the door completely seals the cooking cavity, the unit is automatically turned off. If leakage occurs, the radiation forms a narrow beam emitted around the periphery of the door. If the interlock completely fails, the emitted radiation with the door open forms a directional but broad beam. Finally, the time of exposure is a critical consideration. As noted earlier, any biological damage from the microwave radiation is a result of the heating of tissue; hence, the length of time of exposure is an important factor in the extent of damage.

Although ovens in good operating condition with relatively low leakage levels do not present a direct hazard to personnel, an indirect hazard could exist to individuals wearing cardiac pacemakers. Potentially, radiation from a microwave oven can disrupt the normal rhythm of a pacemaker. Cardiac pacemakers may be affected by microwave levels on the order of one one-hundredth (1/100) of the normal maximum permissible exposure of 10 milliwatts per square centimeter (mW/cm^2). Individuals utilizing pacemakers should be cautioned not to enter areas where microwave ovens are in use. Hospital personnel utilizing microwave ovens on mobile food carts should recognize the danger involved in using these ovens in the vicinity of patients with pacemakers.

Although without suitable instrumentation it is not possible to assess the exact amount of radiation a microwave cooking oven is leaking, there are certain visual checks which can be made. As almost all instances of leakage are around the door area, the following visual checks are good precautionary measures:

- Check for loose or bent door hinges, screws missing from hinges.
- Sprung, warped, or misaligned doors.
- Faulty interlocks; for example, oven shoud not be operable with door open or slightly ajar.
- Worn, missing, or damaged seals around the door or viewing area.
- Check for pitting and burnt spots around the periphery of the door closure area. This is usually caused by arcing as a result of grease buildup around the door. Ovens should be checked at frequent intervals to eliminate this arcing, which causes an increase in leakage levels.
- Check to see if personnel are using metal or aluminum foil cooking vessels, as reflections from metal objects can also increase leakage.
- Check to see that ovens are not being operated empty. If for any reason the oven must be operated without food, to check its operation, such as interlocks, a small bowl or beaker of water should be used to simulate a normal load.
- Appropriate operating signs should be posted.

ELECTROMAGNETIC INTERFERENCE

In addition to potential damage to electronic prosthetic devices, diagnostic medical equipment, such as electroencephalograms, electrocardiograms, electromyographs, etc., can be affected by microwave radiation. This effect is known as electromagnetic interference (EMI). EMI occurs when signals generated by one or more electronic or electromechanical devices adversely affect the operation of other electron devices. The offending signals can produce a variety of intended radiation, such as radio, television and radar transmission, or unintended signals generated by internal combustion engine ignition systems, electric razors, electric motors, electromechanical relays, etc. The characteristic of electronic equipment that permits undesirable responses when subjected to EMI is referred to as susceptibility, and the prevention techniques are referred to as the science of electromagnetic compatibility (EMC). The level of field strength of EMI is usually measured in terms of the E field vector, or in volts per meter (v/m). However, the H field vector in amperes per meter is also often measured. In order to have some idea of the field strength levels of some common emitters, a representative list is included below.

To determine field strength from power density, the conversion is: $E = 61.4\, P_d$, where E = field strength in V/m, and P_d = power density in milliwatts/square centimeter (mW/cm^2). This comes from $P_d = E^2/3770$ solved for

E. Note that the personnel hazard limit is 195 V/m free space equivalent, and that EMI levels as low as 0.6 V/m have caused pacemaker malfunctions. Examples of possibly damaging field strengths include the following:

1. Field strength a quarter mile from a 50-kilowatt broadcast station = 150 V/m (6 mW/cm²).
2. Walking across the beam of a ship's radar = 300 V/m (25 mW/cm²).
3. Beneath the dipole antenna of a 1 kW amateur radio station = 30 V/m (.25 mW/cm²).
4. Close proximity to a VHF police or taxi transmitter = 35 V/m (.32 mW/cm²).
5. Close proximity to a 5-watt citizen band transmitter = 20 V/m (.11 mW/cm²).
6. Beneath a high voltage power line = 3000 V/m (2400, mW/cm²).

Cardiac pacemakers can be categorized into two general types. They are the demand (or synchronous) type and the continuous (or asynchronous) type. Pacemakers are usually implanted in the right rib cage, with catheters containing the signal wires routed intraveneously to the right ventricles of the heart. The demand type of pacemaker is the more popular type for patient implant because it provides pacing only when the normal depolarization signals of the heart are not sensed. When this occurs, the pacemaker then delivers a fixed rate electrical pulse to stimulate the heart's ventricular muscle. The stimulus is applied until the pacemaker circuitry again senses the normal heart function. The older "continuous type" of pacemaker which provides continuous heart stimulation is still used to some extent where there is complete or chronic atrioventricular (AV) heart block. This type is less susceptible to EMI than the demand type because the circuitry necessary to sense the level heart depolarization impulses is not required. Demand type pacemakers usually have interference circuitry built in to detect interference of a continuous wave nature. When this occurs, the pacer will revert to fixed rate pacing. Fixed rate pacing will not seriously affect the user, provided the rate is sufficiently high to sustain the user's activity without producing competitive rhythms (which may cause fibrillation). The most serious interference to the demand type pacer is that which misguides the pacer into its inhibit mode, thereby denying the heart proper stimulus. Should the heart be in AV block when the inhibit occurs, death could result. Note that placing the pacemaker into a shielded package can decrease susceptibility somewhat, but it is generally felt that the combination of the pacer and catheter acts as a receiving system. In band or signals that closely approximate the intended signals carried by the system (the detected *R* wave for instance), are readily picked up by the receiving combination. Filtering is not effective for this type of EMI because the intended signal would also be affected. RF frequencies above 3–4 GHz do not pose a problem, even for particularly sensitive pacers, as the body (or skin) provides considerable attenuation to these higher frequencies.

Electronic medical equipment such as EEG, EKG, and EMG, oscillographic displays, recorders, etc., because of their circuit sensitivities, types of installation, and the length of sensors normally required, are also susceptible to common EMI. EMI to this type of equipment can cause erroneous indications which complicate the diagnostic problem and may even result in equipment malfunction through component failure (transistor damage, etc.). The EMI problems associated with this type of equipment is often further compounded by a need for portability and flexibility. Since most of the above equipment can be thought of as receiving combinations of one type or another, extraneous EMI levels in the millivolt per meter (.001 V/m) range could affect proper operation. Medical electrical/electronic equipment itself can be a source of EMI. RF and microwave diathermy, electrocoagulation units, microwave blood warmers and some monitor read-out devices such as digital voltmeters and frequency counters are all potential sources of harmful interference. Electromechanical devices such as motors, pumps, relays, and switches, and even alternating current power cables can also be added to the list. Suggested readings at the end of this chapter provide guidelines to remedies for these situations.

REFERENCES

1 Barnes, F. S. and C. H. Hu. "Model for Some Nonthermal Effects of Radio and Microwave Fields on Biological Membranes," *IEEE Trans. Microwave Theory Tech.*, MIT–25, 742–746 (1977).

2 Cain, C. A. "Theoretical Basis for Microwave and RF Field Effects on Excitable Cellular Membranes," *IEEE Trans. Microwave Theory Tech.*, MIT–28, 142–147 (1980).

3 Chou, C. K. and A. W. Guy. "Effect of 2450 MH_2 Microwave Fields on Peripheral Nerves," *Dig. Tech. Papers* (IEEE Int. Microwave Symp., Boulder, CO), 318–320 (1973).

4 Clapman, R. M. and C. A. Cain. "Absence of Heart-Rate Effects in Isolated Frog Heart Irradiated with Pulse Modulated Microwave Energy," *J. Microwave Power*, 20:411–419 (1975).

5 Courtney, K. R., J. C. Lin, A. W. Guy and C. K. Chou. "Microwave Effect on Rabbit Superior Cervical Ganglion," *IEEE Trans. Microwave Theory Tech.*, MIT–23, 809–813 (1975).

6 Frey, A. H. and E. Seifert. "Pulse Modulated UHF Energy Illumination of the Heart Associated with Change in Heart Rate," *Life Sci.*, 7:505–512 (1968).

7 Pickard, W. H. and F. J. Rosenbaum. "Biological Effects of Microwave at the Membrane Level," *Math. Biosci.*, 39:235–253 (1978).

4/ULTRAVIOLET RADIATION

PROPERTIES OF LIGHT AND ULTRAVIOLET RADIATION

Ultraviolet radiation is that portion of the electromagnetic spectrum described by wavelengths between 4 to 400 μm. Table 4.1 lists the three ultraviolet spectrum bands classified by the Commission Internationale de L'Éclairage (CIE), Paris. This International Commission on Illumination classified "UV-C" between 200 and 280 nm due to its limited biological significance, since radiation beyond this region is absorbed in the air with the production of ozone. Ozone is produced mainly at wavelengths less than 2200 nm.

There are two general sources of ultraviolet radiation, namely, natural and artificial. The most important natural source of ultraviolet is the sun. Artificial sources come from a variety of electric arcs and lamps.

The sun is essentially an incandescent mass having a surface temperature of 6000°K. Its internal temperature has been estimated to be about 20 million degrees. The intensity of the solar radiation outside of the earth atmosphere at the earth's mean solar distance, 148.6 million kilometers, is called the solar constant. It has an averaged value of 2 cal/cm²/min. About two-thirds of this energy reaches the earth's surface, the remainder being scattered, reflected, or absorbed by the atmosphere. The energy distribution in the solar spectrum observed at the earth's mean distance from the sun indicates about half the sun's energy is in the visible region, about 40% is in the infrared region, and about 10% in the ultraviolet region. Wavelengths shorter than 295 nm are absorbed by the atmosphere so that there is very little of it that reaches the surface of the earth.

Artificial sources either produce ultraviolet as a byproduct, or are designed to generate ultraviolet to capitalize on its properties. The amount of radiation which lies in the ultraviolet region varies with the nature of the arc. The main classes of arcs are open and closed types. Open arcs are discharges taking place between two electrodes exposed to the atmosphere. Ultraviolet radiation

Table 4.1. Ultraviolet band spectrum.

UV-A	315–400 nm
UV-B	280–315 nm
UV-C	100–280 nm

produced by open arcs is a byproduct and exists where the following operations occur: arc welding, plasma torch, glass blowing, hot metal operation,
and photoelectric scanning.

Enclosed arcs are discharges through a gas or vapor contained in an enclosure, usually in quartz. Examples of the gases used are mercury, helium,
neon, krypton, xenon and argon. Arcs such as these are ultraviolet radiation
generators. In the area of germicide, artificial sources are used in hospital,
school, biological laboratories and in industry. Other examples include chemical synthesis and analysis; food, water, and air sterilization; photoengraving;
vitamin production; ink drying in canning industry; and crime detection. Examples of disinfection operations are given by Cheremisinoff et al. (1981).
Figure 4.1 shows the electromagnetic spectrum for light.

WAVE THEORY OF LIGHT

The properties of light may be described by wave theory which is general
for all types of wave phenomena. The ray theory of light is an approximation
of wave theory and is valid only when the size of the light beam and the obstacles with which it interacts are much greater than the wavelength of light.

Light is a propagating wave which falls within the electromagnetic spectrum. The basic difference between the various types of electromagnetic waves
is their differing wavelengths. All electromagnetic radiation travels through a
vacuum with the speed of light, c. That is, for all electromagnetic radiation:

$$\lambda = \frac{c}{f} \tag{4.1}$$

where

f = frequency of radiation
λ = wavelength

Radiation in the range of wavelengths between that of visible light (7×10^{-7} m) and radar waves is infrared. Infrared radiation is readily absorbed by
most materials. The wavelengths of the visible portion of electromagnetic ra-
diation are in the range of about 4×10^{-7} to 7×10^{-7} m. The various wave-

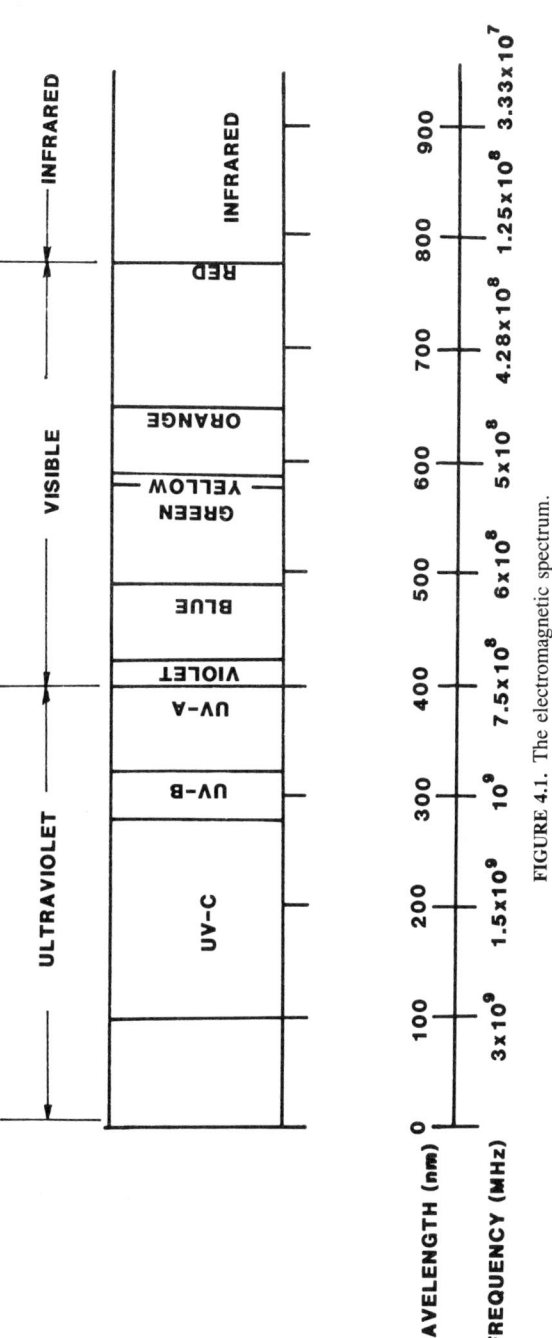

FIGURE 4.1. The electromagnetic spectrum.

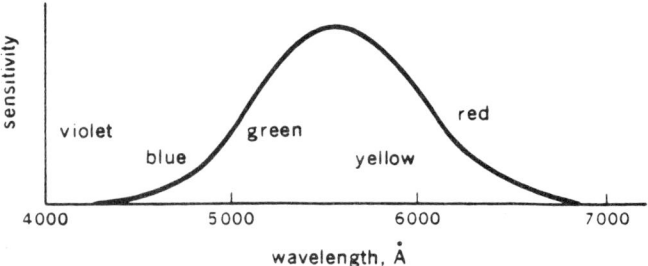

FIGURE 4.2. Color sensitivity curve for the human eye.

length regions in this range are classified by colors. The sensitivity of the human eye to wavelengths in this region is shown in Figure 4.2. Ultraviolet light is radiation with λ shorter than visible violet light but still longer than approximately 100 Å.

The field structure of a ray of light can be represented as shown in Figure 4.3. Using the coordinate system defined in this figure, the ray travels along the z-axis in the positive direction. That is, it physically occupies only the one-dimensional line (z-axis) and does not extend to the other side.

Electric (E) and magnetic fields (H) are denoted by vectors along the x and y axes, respectively. The magnitude of the vector reflects the field strength at a point on the ray. It does not represent a distribution that extends to one side. Although Figure 4.3 does show the vectors extending to the side, this is not a

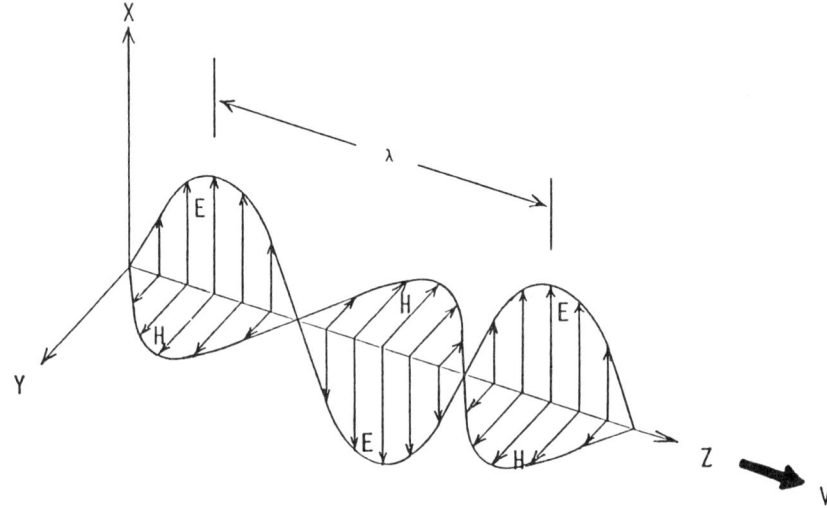

FIGURE 4.3. Field structure of a ray of light.

property of a ray of light, but rather the restriction imposed by a two-dimensional drawing. Other fields in fact do exist on either side of the ray; however, they must be represented by additional rays positioned adjacent to the one shown in Figure 4.3 but are not necessarily parallel to it.

The *E* and *H* fields are normal to each other, and the direction of propagation (where *E* is in the x-direction, *H* in the y-direction, and propagation is in the z-direction).

The direction of the *E*-field is referred to as polarization. Figure 4.3 shows an x-polarization, which is arbitrarily chosen. We may also denote a y-polarization, or any in a direction within the x-y plane.

By definition, a plane electromagnetic wave is a wave in which *E* and *H* are constant at a given instant at any plane perpendicular to the direction of propagation. There are two well-known facts about such a wave. First, the energy stored in a unit volume of space where the electric field is *E* is $\epsilon_0 E^2/2$, where ϵ_0 is the permittivity constant. Second, the energy stored in a unit volume of space where the magnetic field is *H* is $H^2/2\mu_0$, where μ_0 is the permeability constant. We can evaluate the average energy per unit volume of a plane electromagnetic wave from these two facts. Consider an imaginary box of unit cross section that is one wavelength long, through which the wave passes (see Figure 4.4). The energy contained in this volume reflects the average energy of the wave since the box is one wavelength long. At time $t = 0$:

$$E_x = E_{0x} \sin \frac{2\pi f Z}{\nu} \tag{4.2A}$$

$$H_y = H_{0y} \sin \frac{2\pi Z}{\lambda} \tag{4.2B}$$

where *f* is frequency and ν is the speed of the crest of the wave.

The energy stored in a thin slab of width ΔZ is

$$1/2 \left(\epsilon_0 E_x^2 + \frac{H_y^2}{\mu_0} \right) \Delta Z$$

For a unit cross section, ΔZ is also the volume. Integrating this expression over the limits of $Z = 0$ to $Z = \lambda$ provides the energy contained in the box:

$$\Sigma = \frac{\lambda}{4} \left(\epsilon_p E_{0x}^2 + \frac{H_{0y}^2}{\mu_0} \right) \tag{4.3}$$

The magnitudes of the *E* and *H* fields are related to each other by the charac-

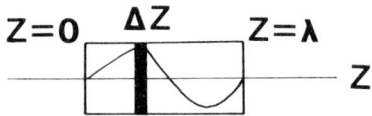

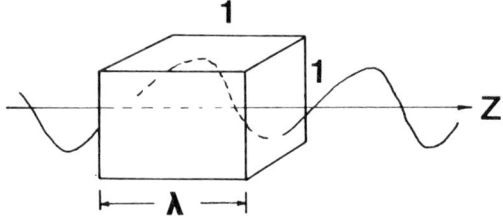

FIGURE 4.4. System defining energy density.

teristic impedance η of the medium through which they travel:

$$E = \eta H \tag{4.4}$$

where η is an intrinsic characteristic of the material ($= 1/\sqrt{\epsilon_0 \mu_0}$). Dividing Equation (4.3) by λ gives the energy density of the wave:

$$\hat{\Sigma} = 1/2\epsilon_0 E_{0x}^2 \tag{4.5}$$

The intensity of the wave (i.e., the energy incident per unit time on unit area of surface perpendicular to the direction of propagation) is

$$I = \frac{\text{Energy Incident}}{\text{(area) (time)}} = 1/2c\epsilon_0 E_{0x}^2 \tag{4.6}$$

where c is the velocity of the wave. Equation (4.6) states that intensity is proportional to the square of the amplitude of the wave. Equation (4.6) may also be stated as:

$$I = 1/2 \, \frac{E_{0x}H_{0y}}{\mu_0} \tag{4.7}$$

It should be noted that all light is polarized. That is, at any instant in time, any ray of light is polarized in a single direction. Furthermore, two or more polarizations do not exist independently at a given point in space and time, but rather add vectorially to produce a single polarization. In reality, most sources emit randomly polarized light (i.e., its direction of polarization fluctuates ran-

domly in time). A polaroid filter will absorb light polarized in one direction
and transmit light polarized in another to provide light of a specific polarization. An example of this is the lens of a pair of polaroid sunglasses.

Distinction should be made between spherical and plane waves. Consider an
ideal point source s which transmits waves of wavelength λ in all directions.
A cross section of the wavefront diagram is shown in Figure 4.5. The motion
of the wavefronts is denoted by rays radiating out from the source. The rays indicate the intensity of the wave. For a steady flow of energy, the intensities at
distances r_1 and r_2 from the source can be related by realizing that the energies
flowing through the spherical shells at these radii are equal:

$$I_1(4\pi r_1^2) = I_2(4\pi r_2^2) \tag{4.8}$$

or

$$\frac{I_1}{I_2} = \frac{r_2^2}{r_1^2} \tag{4.9}$$

That is, the intensity of a spherical wave decreases as the inverse square of the
radius.

For a very large radius, $I_1 \cong I_2$. This is true for $r_1 = r_2 - \delta$ and $\delta << r_2$. In this case, I is nearly independent of position and the rays become essentially parallel. This is the definition of a plane wave (i.e., plane waves are
spherical waves for which $R \rightarrow \infty$). The intensity of a plane wave is constant

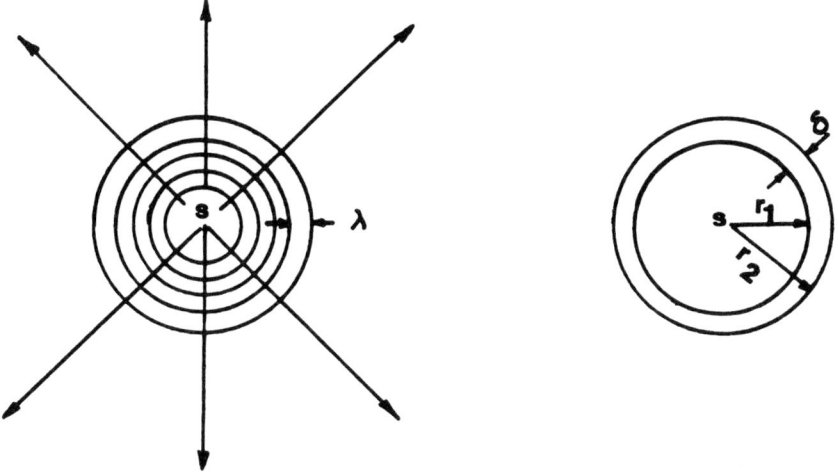

FIGURE 4.5. Cross section of wavefront diagram.

at any point in space. It should be noted that the distinction between spherical and plane waves is based on an idealized source which transmits a perfect sinusoidal-type wave. Light waves are not exactly of this form; however, they can be approximated by the sum of pure sinusoidal waves (Fourier series).

Discussions to this point have described the properties of a light wave in terms of it being frozen in time and space. Consider now the wave to travel along the Z axis with velocity v. The velocity of the light wave through a medium of index of refraction n is

$$v = \frac{c}{n} \tag{4.10}$$

where c = vacuum velocity of light (2.997925×10^8 m/s).

The traveling light wave has a wavelength λ, defined as the distance between successive peaks in the E field. For a definite medium:

$$\lambda = \frac{\lambda_0}{n} \tag{4.11}$$

where λ_0 = vacuum wavelength. If we consider ourselves as stationary observers on the Z-axis, the E field peaks are seen to pass by having frequency f defined as

$$f = \frac{v}{\lambda} = \frac{c}{\lambda_0} \tag{4.12}$$

Equation (4.12) shows that although the velocity and wavelength of light vary from one medium to the next, the wave's frequency is preserved.

We now add another dimension to the light beam. Instead of considering it as having an infinitesimal width, assume the beam to have a finite thickness. If we travel along with the crest of the wave (i.e., ride on the E-field peak of the ray), the peak is observed to extend to either side. As noted earlier, the wave is referred to as a wavefront. The wavefront defines the ray which is always perpendicular to it. Examples are shown in Figure 4.6. Figure 4.6(A) shows the wavefronts and rays in a collimated beam of light. Figure 4.6(B) shows the wavefronts and rays in a focused beam of light. In the latter example, the wavefronts are spherical surfaces whose radii of curvature decrease to zero as they propagate to the focus, and then increase on the other side. Hence, the rays are not parallel.

The principle of reflection is illustrated by the ray diagram in Figure 4.7. Angle Θ_i is the angle of incidence, defined as that angle between the ray and

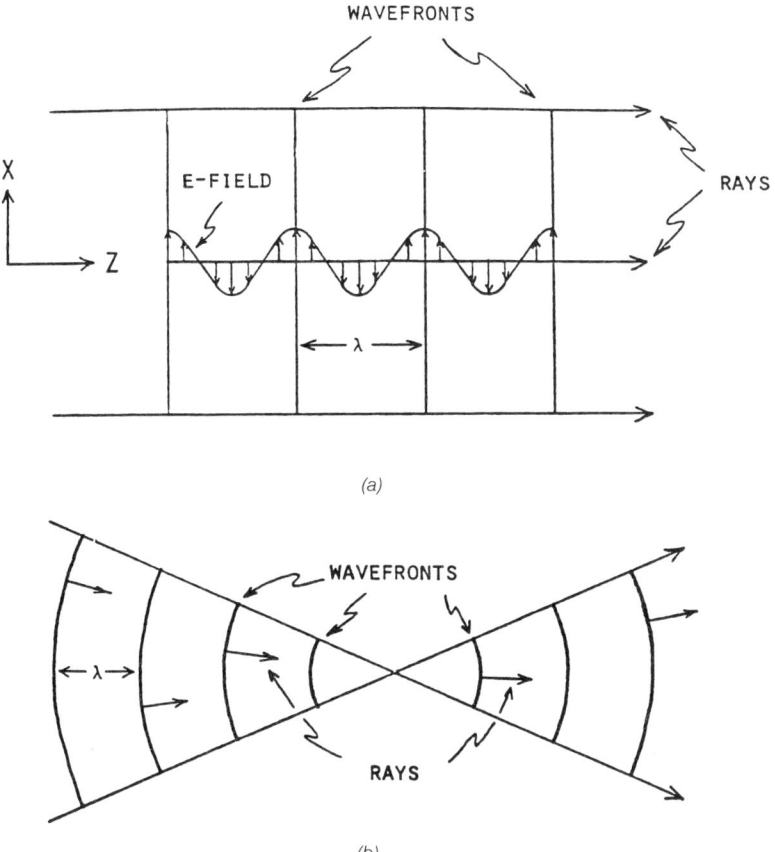

FIGURE 4.6. (a) collimated light beam; (b) focused light beam.

the normal to the reflecting surface. The velocity vector of the ray can be
resolved into two components: one parallel to the surface and the other perpendicular to it. The velocity component perpendicular to the surface is
reversed upon reflection while the other component remains unchanged. The
reflected wave also appears as a plane wave. The important observation from
Figure 4.7 is that the angle of incidence Θ_i equals the angle of reflection Θ_r.
If attention is directed to an infinitesimal point on a large surface, then at that
point of reference the observer views an infinite flat plane. Consequently, the
concept of $\Theta_i = \Theta_r$ applies under all conditions regardless of the roughness
or geometry of the overall surface. Also, note that $\Theta_i = \Theta_r$ for individual
rays, independently of whether the wavelength is plane or spherical.

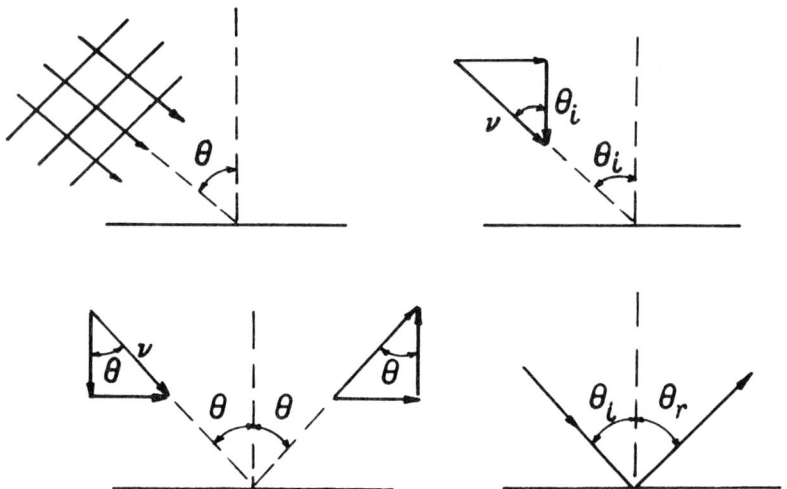

FIGURE 4.7. Illustrates light reflection.

The principle of reflection is practically applied through the use of mirrors. A plane mirror for example, is a limiting form of a spherical mirror (i.e., a spherical mirror for which the radius of curvature approaches infinity).

For a concave mirror (see Figure 4.8) two rays are observed from the tip of the object O. Ray 1 intersects the center of the mirror and is reflected with $\Theta_i = \Theta_4$. Ray 2 passes through point c, which is the centerpoint of the concave mirror. Note that all radii are perpendicular to the surface of the sphere and hence, ray 2 is reflected back on itself. To the human eye, rays A and B appear to originate from point O'. Hence, the image of O is seen at O'. The object distance s and the image distance s' can be related through single geometric relations:

$$\tan \Theta_r = \frac{O'}{s'} \ , \ \tan \Theta_i = \frac{O}{s}$$

where O' and O denote the lengths of image O' and object O, respectively.

From the law of reflection it is easy to show that

$$\frac{O'}{O} = \frac{R - s'}{s - R} \tag{4.13}$$

and further

$$\frac{1}{s} + \frac{1}{s'} = \frac{2}{R} \tag{4.14}$$

Equation (4.14) relates object distance s to image distance s' through the mirror's radius of curvature. Note that a clear image is only obtained if the rays are restricted to the central region of the mirror.

Recall that when an object is far removed from the mirror (i.e., $1/s \rightarrow O$), the image will appear at the mirror's focal point. In terms of the focal length for a concave mirror ($F = R/2$):

$$\frac{1}{s} + \frac{1}{s'} = \frac{1}{F} \tag{4.15}$$

Because of the symmetry between s and s', the direction of light rays can be reversed. For example, light originating from a source positioned at the focal point would be reflected as parallel light (i.e., as a plane wave). This offers a convenient way for obtaining collimated light.

We now review the phenomenon of refraction. When a beam of light passes

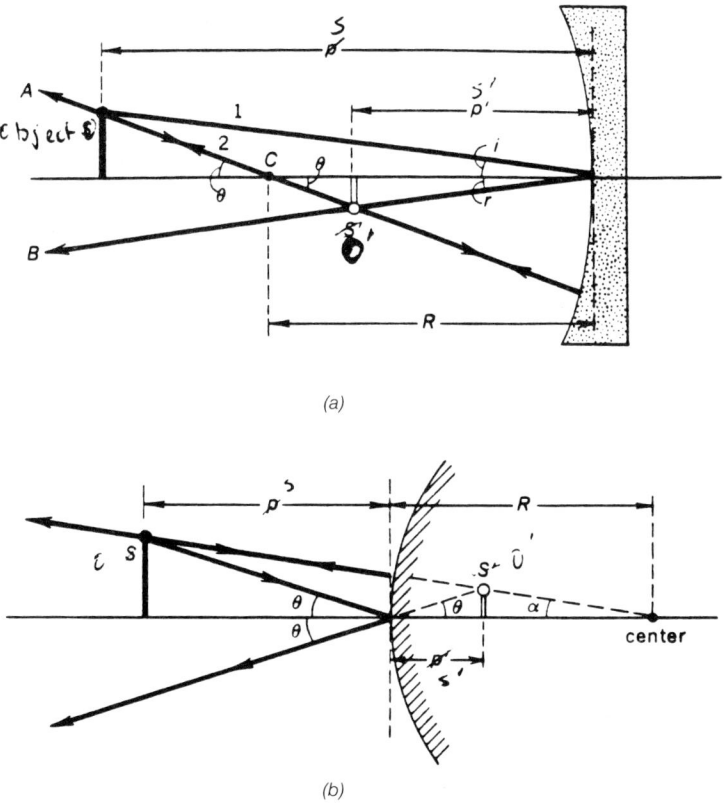

(a)

(b)

FIGURE 4.8. (a) laws of reflection for a concave mirror; (b) reflection from a convex mirror.

from one medium to another and if the speed of light is different in the two materials, the light beam is refracted (see Figure 4.9). For $\nu_2 < \nu_1$, the wavefronts are closer together in material 2 of the figure. In the example shown, the frequency of wave crests passing points A and B is the same. The ratio of distance between wavefronts is

$$\frac{\lambda_1}{\lambda_2} = \frac{\nu_1}{\nu_2} \qquad (4.16)$$

From an algebraic derivation, it is easy to develop Snell's law

$$\frac{\sin \Theta_1}{\sin \Theta_2} = \frac{c}{\nu_2} \equiv n \qquad (4.17)$$

where the first medium is taken to be a vacuum, and n_a is the absolute index of refraction of the second material. Typical values of n are given in Table 4.2. The relative index of refraction is defined as

$$\frac{\sin \Theta_1}{\sin \Theta_2} = \frac{n_2}{n_1} = n \qquad (4.18)$$

n_1, n_2 are the absolute refractive indices of mediums 1 and 2, respectively.

An important law is Fermat's principle of least time, which may be stated as follows: A beam of light which originates from point A and travels to point B

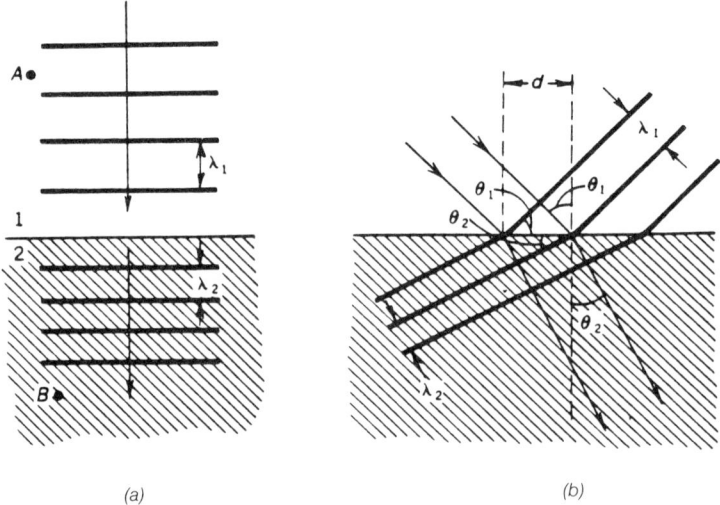

(a)   (b)

FIGURE 4.9. Refraction caused by light passing through different media.

Table 4.2. Typical values for absolute index of refraction.

Material	n_a	Material	n_a
Air (STP)	1.003	Sodium Chloride	1.53
Water	1.33	Carbon Disulfide	1.63
Ethanol	1.36	Diamond	2.42
Benzene	1.50	Fused Quartz	1.46

will follow the path between the two points requiring the shorter traveling time. This principle can be used to derive the laws of reflection and refraction and provides a fundamental means of evaluating the path taken by light beams.

A limiting case of importance is that of total internal reflection. When a light beam travels from a material of high index of refraction to one of lower index, it is possible to have total internal reflection. The situation is illustrated in Figure 4.10. For a critical angle of incidence, Θ_c, the angle of reflection is 90° and hence, for any $\Theta_i > \Theta_c$, the beam is entirely reflected. The critical angle can be evaluated from Snell's law (see Equation 4.17):

$$\sin \Theta_c = \frac{n_1}{n_2} \ (\text{for } n_1 > n_2) \tag{4.19}$$

(note: $\sin \Theta_2 = \sin 90° = 1$).

Total internal reflection allows light to be transmitted around corners, for example, by using a rod of transparent material or a bundle of fibers (i.e., fiber optics).

All common waves obey the superposition principle which states that when several wave disturbances simultaneously meet at a point, the resultant disturbance is the vector sum of the separate disturbances. It is possible to represent any periodic function by a Fourier series.

A review of the mathematical principles is worthwhile. Consider two waves of the same amplitude but whose phase differs by ϕ. Expressions approximating these signals are

$$y_B = y_0 \sin \omega t$$
$$\tag{4.20}$$
$$y_A = y_0 \sin (\omega t - \phi)$$

where

$$\phi = 2\pi \frac{\delta}{\lambda} \tag{4.21}$$

δ is defined as the path difference between the two waves.

The sum of the two waveforms is

$$y = y_0 [\sin \omega t + \sin (\omega t - \phi)] \qquad (4.22)$$

And by trigonometric simplification

$$y = 2y_0 \cos (1/2\phi) \sin (\omega t - 1/2\phi) \qquad (4.23)$$

The resultant wave is sinusoidal, having the same frequency ω. The amplitude changes by a factor $2 \cos 1/2\phi$. For phase angle $\phi = 0, 2\pi, 4\pi, \ldots$ the amplitude of the oscillation is $2y_0$. Also, when $\phi = \pi, 3\pi, 5\pi, \ldots$ the $\cos 1/2\phi$ is zero, which is the criterion of destructive interference. The intensity of the resultant wave is

$$I = 4y_0^2 \cos^2 (1/2\phi) \sin^2 (\omega t - 1/2\phi) \qquad (4.24)$$

And the average intensity is

$$I_{av} = I_0 \cos^2 1/2\phi \qquad (4.25)$$

(This is based on the average of $\sin^2 \Theta = 1/2$, and noting $I_0 = 2y_0^2$.)

We may now state one of the most fundamental aspects of electromagnetic radiation and optical interference phenomena: At any point in space, all electric and magnetic fields add vectorially to form a single electric and magnetic field vector pair. If two E-fields have different polarizations, their sum cannot cancel. For E fields of the same polarization, destructive interference results when their magnitudes are the same and their directions are opposite. This is illustrated in Figure 4.11 showing that only E fields, or field components, of like polarization interact to produce interference.

Note that if the magnitudes of the E fields are not equal, partial cancellation still occurs [Figure 4.11(c)]. The intensity, however, does not fully cancel at the interference modes in this example.

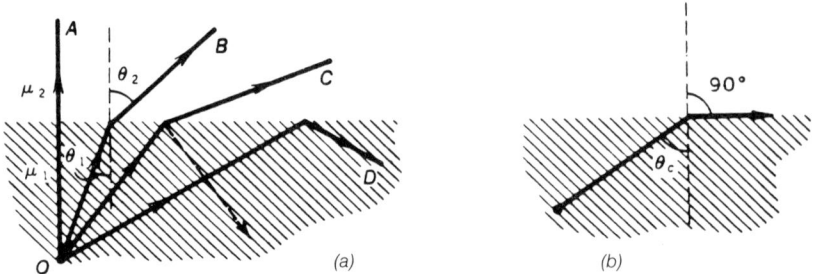

FIGURE 4.10. Illustrates total internal reflection.

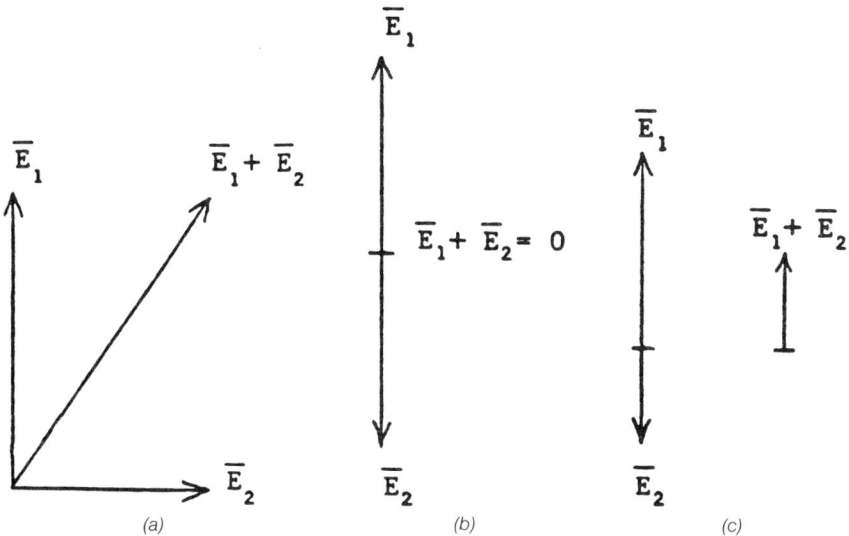

FIGURE 4.11. Shows interaction (addition) of electric field vectors.

BIOLOGICAL EFFECTS AND HAZARD ANALYSIS

For the ultraviolet spectral region of 200 to 315 nm, total irradiance incident on unprotected skin or eyes, based on either measurement data or on output data, can be determined as follows (based on NIOSH Standards—*Occupational Exposure to Ultraviolet Radiation*, HSM 73–11009):

1. If the ultraviolet energy is from a narrow-band or monochromatic source, permissible level for a daily 8-hour exposure.
2. If the ultraviolet energy is from a broad-band source, the effective irradiance (I_{eff}) relative to a 270 nm monochromatic source shall be calculated from the formula below. From I_{eff}, the permissible exposure time in seconds for unprotected skin or eyes shall be computed by dividing 0.003 J/cm², the permissible dose of 270 nm radiation, by I_{eff} in W/cm².

$$I_{eff} = \Sigma I_\lambda S_\lambda \Delta_\lambda \qquad (4.26)$$

where

I_{eff} = effective irradiance relative to a monochromatic source at 270 nm

I_λ = spectral irradiance in W/cm²/nm

S_λ = relative spectral effectiveness (dimensionless)

Δ_λ = band width in nm.

Table 4.3. Maximum permissible exposure times for selected values of I_{eff}.

Duration of exposure per day	Effective irradiance, I_{eff} (μW/cm^2)
8 hrs	0.1
4 hrs	0.2
2 hrs	0.4
1 hr	0.8
30 min	1.7
15 min	3.3
10 min	5.0
5 min	10.0
1 min	50.0
30 sec	100.0

Table 4.3 lists permissible exposure times corresponding to selected values of I_{eff} in μW/cm^2 (see also Table 4.4).

If radiation intensity from a point source is known at some distance from the worker, for example, from measurement at another point or from output data at a known distance from the ultraviolet source, attenuation of radiation from that point to the worker can be calculated from the principle that radiation decreases with the square of the distance it must travel. For example, an object 3 feet away from a radiation source receives 1/9 the energy of an object 1 foot

Table 4.4. Total permissible 8-hour doses and relative spectral effectiveness of selected monochromatic wavelengths.

Wavelength (nm)	Permissible 8-hour dose (mJ/cm^2)	Relative spectral effectiveness ($S\lambda$)
200	100.0	0.03
210	40.0	0.075
220	25.0	0.12
230	16.0	0.19
240	10.0	0.30
250	7.0	0.43
254	6.0	0.50
260	4.6	0.65
270	3.0	1.00
280	3.4	0.88
290	4.7	0.64
300	10.0	0.30
305	50.0	0.06
310	200.0	0.015
315	1000.0	0.003

away. This assumption is conservative since ultraviolet radiation, especially at very low wavelengths, may be absorbed by some components of the atmosphere. Where information on atmospheric absorption of ultraviolet radiation is known, further correction may be applied. The calculation of intensity of radiation at any given point by use of the inverse square formula does not take into consideration reflected energy.

It should be noted that significant non-occupational exposure to ultraviolet radiation can occur from exposure to sunlight, particularly during the sumer months.

The National Institute for Occupational Safety and Health (NIOSH) recommends that occupational exposure to ultraviolet energy in the workplace be controlled. Ultraviolet radiation (Ultraviolet energy) is defined as that portion of the electromagnetic spectrum described by wavelengths from 200 to 2000 nm. Adherence to the recommended standards will prevent occupational injury from ultraviolet radiation, that is, will prevent adverse acute and chronic cutaneous and ocular changes precipitated or aggravated by occupational exposure to ultraviolet radiation.

Sufficient technology exists to prevent adverse effects on workers, but technology to measure ultraviolet energy for compliance with the recommended standard is not sophisticated. Consequently, special work practices are recommended for control of exposure in cases where sufficient measurement or emission data are not available.

The exposure standards as outlined by NIOSH are as follows:

- For the ultraviolet spectral region of 315 to 400 nm, total irradiance incident on unprotected skin or eyes, based on either measurement data or on output data, shall not exceed 1.0 mW/cm² for periods greater than 1000 seconds, and for exposure times of 1000 seconds or less, the total radiant energy shall not exceed 1000 mW . sec/cm² (1.0 J/cm²).

- For the ultraviolet spectral region of 200 to 315 nm, total irradiance incident on unprotected skin or eyes, based on either measurement data or on output data, shall not exceed levels provided in Table 4.3 and Table 4.4.

OSHA Safety and Health Standards 1910.133, 1915.81, 1916.81 and 1926.102 provide requirements for eye and face protection which may be applied to hazards other than those associated with welding operations. The final section of 1910.133 states: "Design, construction, testing and use of device for eye and face protection shall be in accordance with American National Standard for Occupational and Education Eye and Face Protection, Z 87-1-1968." This may be considered a national consensus standard, enforceable under the Occupational Safety and Health Act of 1970.

CONTROL AND PREVENTION PRACTICES

Since the eye and skin are the most sensitive organs, coupled with the inability of human senses to detect the presence of ultraviolet radiation, there is a need for effective control measures for equipment capable of producing ultraviolet radiation. Furthermore, outdoor tasks where solar radiation exists can be considered an occupational exposure.

There are a variety of control measures and control techniques that can be implemented in the workplace.

The first of these is isolation. Isolation of an ultraviolet source can be provided by distance, enclosure, or shieldings.

The principal that radiation intensity decreases with the square of the distance it must travel, can be applied to reduce potential exposure by relocating either the source of radiation or the workforce. Provided that the reflected radiation is not a factor, the Inverse Square Law relation is

$$l_1(D_1)^2 = l_2(D_2)^2 \tag{4.27}$$

This demonstrates that radiation intensity is reduced by a factor of 4 for each doubling of the distance between the source and receptor.

Total enclosure of the operation is a common control practice for intense UV sources such as those used in plastics curing and carbon arc processes. Interlocks are frequently included to disrupt the power supply in the event that the enclosure is opened. Welding screens are required by the OSHA Safety and Health Standards for "protection from arc welding rays." Where the work permits, the welder should be enclosed in an individual booth painted with a finish of low reflectivity such as zinc oxide (an important factor for absorbing ultraviolet radiations) and lamp black. Another possibility is for the welder to be enclosed by noncombustible screens similarly painted. Booths and screens permit circulation of air at floor level. Workers or other persons adjacent to the welding areas should be protected from the rays by noncombustible or flameproof screens or shields and be required to wear appropriate goggles.

Reflections of UV radiation is a concern not only with welding screens but also with other barriers such as walls and ceilings. Since painted walls, ceilings, screens, and barriers can be a significant source of ultraviolet reflection and potential exposure to individuals in surrounding work areas, effective control demands consideration of the UV reflective properties of the surface coatings.

The properties of a paint depend upon the nature and amount of the pigment and the state of its aggregation. The addition of a small amount of colored pigment to a white paint may result in a large decrease in the ultraviolet reflection. The reflectance may decrease with increase in amount of added colored pigment.

The reflection of incident ultraviolet radiation from pigments can range from negligible to more than 90 percent. It is important to note that a material's ability to reflect visible light is no indication of its ability to perform similarly with ultraviolet. Reflection from selected materials at several wavelengths in the ultraviolet range is given in Table 4.5. Note that ordinary white wall plaster has a reflection of 46 percent at 253.7 nm, whereas zinc and titanium oxides, which are equally good reflectors for visible light, reflect only 2.5 percent and 6 percent respectively, at the same wavelength.

One should first investigate the composition of white pigments before making estimates of their reflections. An example which illustrates this point is the difference between the two white pigments zinc oxide and white lead. Although both of the pigments are very good reflectors of visible radiation, zinc oxide reflects only 3 percent of the ultraviolet, whereas white lead reflects about 60 percent. In general, colored pigments are poor reflectors of ultraviolet.

Table 4.6 provides data on the ultraviolet reflectance of a number of dry white pigments over the range of 280 to 320 nm. These data were obtained using unresolved radiation from a cadmium lamp as a source and an S-1 phototube as a detector. The measurements may be assumed to be predominantly at the wavelength 302.4 nm.

In addition to isolation of the UV source, interruption of the radiation via shielding is an engineering control that can be used to alter the nature of the wave before it reaches workers. One example of this approach to control is the

Table 4.5. Reflection of white pigments and selected materials.*

	2537 Å in Percent	2967 Å in Percent	3650 Å in Percent	Visible Light in Percent
Pressed Zinc Oxide	2.5	2.5	4	88
Barytes	65	70	77	86
Titanium Oxide	6	6	31	94
Pressed Magnesium Oxide	77	86	87	93–95
Smoked Magnesium Oxide	93	93	94	95–97
Pressed Calcium Carbonate	78	83	86	96
White Wall Plaster	46	65	76	90
SW White Decotint Paint	33	41	58	79
Kalsomine White Water Paint	12	20	40	70
Albastine White Water Paint	10	14	45	78
White Porcelain Enamel	4.7	5.4	63	80
Flat Black Egyptian Lacquer	5	5	5	5
Five Samples of Wallpaper	13–31	21–40	33–50	55–75

* M. Lucklash, *Applications of Germicidal, Erythermal and Infrared Energy*, New York, D. Van Nostrand Company, 1948, page 383.

Table 4.6. Ultraviolet reflectance of dry white pigments.*

Pigment	Ultraviolet Reflectance Factor in Percent
Lead-free zinc oxide	3
35% leaded zinc oxide	4
Zinc sulfide	6
Titanox B	6
Titanium oxide	7
Titanox C	7
Lithopona	8
Antimony oxide	17
Zirconium oxide (commercial)	41
Diatomaceous silica (Celite 110)	45
Basic sulfate white lead	48
China clay	54
Aluminum oxide	55
Basic carbonate white lead (Dutch process)	62
Aluminum hydroxide	67
Zirconium oxide, C.P.	78
Magnesium carbonate (commercial)	81

* D. F. Wilcock and W. Soller; *Ind. Eng. Cham.* 32, 1446, 1940.

mercury vapor lamp. Light is produced when an electric current is passed through mercury-argon vapor contained in a quartz tube. To filter out short-wave radiation and protect the inner metal parts from oxidation, the lamp is shielded by a hard borosilicate glass envelope. If this protective shield is broken, many lamps will continue to operate for a short time and emit unfiltered UV radiation.

Structural and portable shields for eye and skin protection can be constructed from ordinary window glass for low-intensity UV sources. Specialized glasses are needed for high-intensity UV sources.

Figure 4.12 reports the percent transmission as a function of wavelength for two thicknesses of window glass. As shown, the transmission falls off rapidly at wavelengths below 360 nm. Window glass 1/8 inch in thickness provides adequate protection for the eyes and skin against ultraviolet radiation from ordinary ultraviolet sources. In the case of very intense sources of ultraviolet radiation, it may not be sufficient.

Full protection against 253.7 nm radiation is provided by shields of clear ultraviolet-absorbing plexiglass, ordinary (glass) spectacles, Crookes glass, and similar ultraviolet-absorbing materials. Crown glass, an alkalilime silicate glass (2 mm thick), will significantly reduce exposure hazards. Flint glass, a heavy glass containing lead oxide (2 mm thick), provides essentially complete protection at all wavelengths. Noviol glasses or Polaroid ultraviolet filters can be used where high-intensity ultraviolet is anticipated, as in welding. Figure 4.13 illustrates the protective potential of Noviol glass, where curves A, B, C, D represent different grades of Noviol glass.

Another prevention route is the use of personal protective devices. The Occupational Safety and Health Standards are very specific for welding operations. With regard to selection, the code states:

1. Helmets or hand shields shall be used during all arc welding or arc cutting operations, excluding submerged arc welding. Goggles should also be worn during arc welding or cutting operations to provide protection from injurious rays from adjacent work, and from flying objects. The goggles may have either clear or colored glass, depending upon the amount of exposure to adjacent welding operations. Helpers or attendants shall be provided with proper eye protection.

2. Goggles or other suitable eye protection shall be used during all gas welding or oxygen cutting operations. Spectacles without side

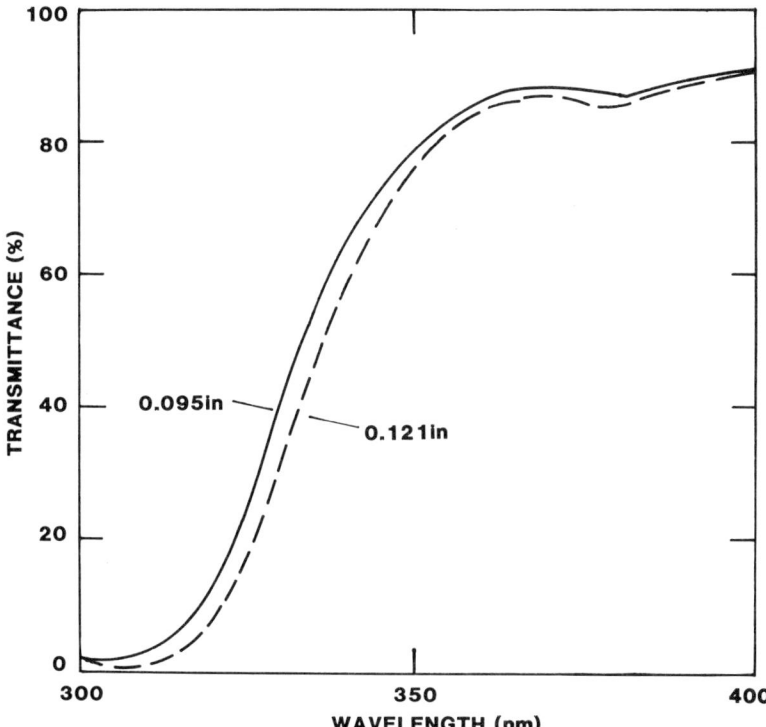

FIGURE 4.12. Transmission for two thicknesses of window glass.

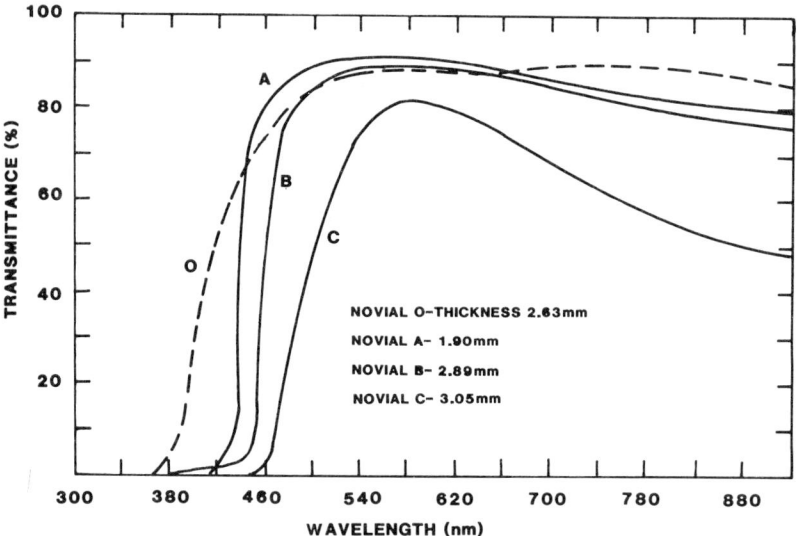

FIGURE 4.13. Transmission through Novial glass.

shields, with suitable filter lenses are permitted for use during gas welding operations on light work, for torch brazing, or for inspection.

3. All operators and attendants of resistance welding or resistance brazing equipment shall use transparent face shields or goggles, depending on the particular job, to protect their faces or eyes, as required.

4. Eye protection in the form of suitable goggles shall be provided where needed for brazing operations not covered in 1, 2 and 3 of the subdivision.

The code further requires the following specifications for protectors.

Helmets and hand shields shall be made of a material which is an insulator for heat and electricity. Helmets, shields and goggles shall be not readily flammable and shall be capable of withstanding sterilization.

1. Helmets and hand shields shall be arranged to protect the face, neck and ears from direct radiant energy from the arc.

2. Helmets shall be provided with filter plates and cover plates designed for easy removal.

3. All parts shall be constructed of a material which will not readily corrode or discolor the skin.

4. Goggles shall be ventilated to prevent fogging of the lenses as much as practicable.

5. Cover lenses or plates should be provided to protect each helmet, hand shield or goggle filter lens or plate.
6. All glass for lenses shall be tempered, substantially free from striae, air bubbles, waves and other flaws. Except when a lens is ground to provide proper optical correction for defective vision, the front and rear surfaces of lenses and windows shall be smooth and parallel.
7. Lenses shall bear some permanent distinctive marking by which the source and shade may be readily identified.

The code contains a guide for selection of the proper shade numbers arranged by type of welding operation.

A useful guide for the selection of eye protection is included in the ANSI Standard, Z 87.1–1968. The ANSI Standard includes a table giving the transmittances and tolerances of various shades of absorptive lenses, filter lenses, and plates (see Table 4.7).

Eye protection filters for glass workers, steel and foundry workers and welders were developed empirically. Optical transmission characteristics are now standardized according to shade, and appropriate shades are specified for particular applications. Although maximum transmittances for ultraviolet and infrared radiation are specified for each shade, the visual transmittance or visual optical density (OD) defines the shade number (S).

Specific protective clothing requirements for welding operations are presented in Section 1910.252 of the code. For most UV exposures, longsleeve garments of densely woven flannelette, poplin, or synthetic fabric will afford sufficient protection. A small cape on the back and sides of the helmets of workers may be necessary during some welding and other operations presenting comparable UV intensity.

When ultraviolet radiation is believed severe enough to burn the skin through these materials, then leathers and/or asbestos should be used for protection. When flame, sparks, or high ultraviolet sources are of concern, flameproof gloves may be necessary.

Individuals exposed to strong solar radiation may require protective clothing such as longsleeved shirts, trousers, skirt, and face and neck protection. The latter may be provided by a broad-brimmed hat, a billed hat or cap, billed hard hat, and face or neck shield. Goggles, spectacles, and skin creams may be considered for some individuals.

Potential exposure to UV radiation should be controlled to the extent possible by engineering means, protective eye wear, and clothing. When shielding the skin is impossible through engineering controls or clothing, sun screens and barrier cream applications should be employed as protective skin applicants.

Ordinary soft paraffin is a good barrier. Barrier creams contain ingredients which will absorb ultraviolet radiation. The benzophenones are the best com-

Table 4.7. Transmittances and tolerances in transmittance of various shades of absorptive lenses, filter lenses, and plates.

Shade Number	Optical Density			Luminous Transmittance			Maximum Infrared Transmittance Percent	Maximum Spectral Transmittance in the Ultraviolet and Violet			
	Maximum	Standard	Minimum	Maximum Percent	Standard Percent	Minimum Percent		313 mu Percent	334 mu Percent	365 mu Percent	405 mu Percent
1.5	0.26	0.214	0.17	67	61.5	55	25	0.2	0.8	25	65
1.7	0.36	0.300	0.26	55	50.1	43	20	0.2	0.7	20	50
2.0	0.54	0.429	0.36	43	37.3	29	15	0.2	0.5	14	35
2.5	0.75	0.643	0.54	29	22.8	18.0	12	0.2	0.3	5	15
3.0	1.07	0.857	0.75	18.0	13.9	8.50	9.0	0.2	0.2	1.0	6
4.0	1.50	1.286	1.07	8.50	5.18	3.16	5.0	0.2	0.2	0.5	1.0
5.0	1.93	1.714	1.50	3.16	1.93	1.18	2.5	0.2	0.2	0.5	0.5
6.0	2.36	2.143	1.93	1.18	0.72	0.44	1.5	0.1	0.1	0.2	0.5
7.0	2.79	2.571	2.36	0.44	0.27	0.164	1.3	0.1	0.1	0.1	0.5
8.0	3.21	3.000	2.79	0.164	0.100	0.061	1.0	0.1	0.1	0.1	0.5
9.0	3.64	3.429	3.21	0.061	0.037	0.023	0.8	0.1	0.1	0.1	0.5
10.0	4.07	3.857	3.64	0.023	0.0139	0.0085	0.6	0.1	0.1	0.1	0.5
11.0	4.50	4.286	4.07	0.0085	0.0052	0.0032	0.5	0.05	0.05	0.05	0.1
12.0	4.93	4.714	4.50	0.0032	0.0019	0.0012	0.5	0.05	0.05	0.05	0.1
13.0	5.36	5.143	4.93	0.0012	0.00072	0.00044	0.4	0.05	0.05	0.05	0.1
14.0	5.79	5.571	5.36	0.00044	0.00027	0.00016	0.3	0.05	0.05	0.05	0.1

pounds for this purpose because of their great absorption capability throughout most of the near and far ultraviolet spectrum.

Sunscreening preparations are usually classified as chemical or physical. Chemical sunscreens include para-aminobenzoic acid and its esters, cinnamates, and benzophenones, all of which act by absorbing radiation so that the energy can be dissipated as radiation of lower energy. The physical agents act as simple physical barriers, reflecting, blocking, or scattering light. They include titanium dioxide, talc, and zinc oxide.

Sunscreen protection from absorbing chemicals depends on maintenance of film thickness with protective power being rapidly lost as the thickness is diminished.

REFERENCES

1 Cheremisinoff, N. P., P. N. Cheremisinoff and R. B. Trattner. *Chemical and Non-chemical Disinfection*. Ann Arbor, Mich.:Ann Arbor Sci. Pub. (1981).

2 Daniels, F., Jr. *Therapeutic Electricity and Ultraviolet Radiation*. S. H. Light, ed. Baltimore, MD:Waverly Press (1967).

3 Food and Drug Administration, *FDA Consumer*, DHEW Publication No. (FDA) 77–8012, U.S. Government Printing Office, 241–191/1 (1977).

4 Freeman, R. C., H. T. Hudson and R. Carneos. "Ultraviolet Wavelength Factors in Solar Radiation and Skin Cancer," *Int. J. Dermatol*, 9 (1970).

5 Hausser, K. W. and W. Vahle, Wiss. Veroff Zern; 101. (1927) translated in *Biological Effects of Ultraviolet Radiation*. F. Burback, ed. NY:Pergamon Press, 101 (1969).

6 Occupational Safety and Health Administration, *OSHA Safety and Health Standards*, 29 CFR 1910, OSHA Publication 2206, Section 1910.252, Washington, DC.

7 National Institute for Occupational Safety and Health, Center for Disease Control, *Nonionizing Radiation-583*, Lesson Plan No. 15, Ultraviolet, Visible and Infrared (May, 1976).

8 National Institute for Occupational Safety and Health, *Occupational Exposure to Ultraviolet Radiation*, U.S. Government Printing Office, Washington, DC (1972).

9 Urback, F., ed. *The Biological Effects of Ultraviolet Radiation*. NY:Pergamon Press (1969).

5 / HEALTH HAZARDS *FROM* LASER RADIATION

APPLICATIONS OF LASER TECHNOLOGY

The use of lasers depends on the properties of the particular light gener-ated, which in turn depends on the type of laser. There are a variety of engi-neering applications, a few examples of which are highlighted in the following paragraphs.

In the communications field, the higher the frequency of a carrier signal, the greater the amount of information that can be impressed upon the carrier. One optical carrier of He-Ne laser frequency (5×10^{14} Hz) could theoreti-cally carry ten million simultaneous phone calls or eight thousand simultane-ous television programs. This ability makes the laser an attractive device for the communications industry. Before communications applications are possi-ble, modulation of the carrier beam must be accomplished. Since the carrier is light, transmission from point-to-point can be stopped by such simple things as fog, rain, dust, or an object passing through the beam. The solution may be in transmission through pipes with mirrors directing the light around bends or through fiber optics. This technology is still in the development stage.

There are several laser tracking and ranging systems presently in use. This application is referred to as LADAR (Laser Detection and Ranging), just as Radio Dection and Ranging is referred to as RADAR. Ranging systems record the time for a signal to travel to and from the target and translate this time to distance. The small divergence of the beam is important because it allows the operator to pinpoint the object for which readings are taken. The Armed Forces have developed a range finder which utilizes this concept.

A collimated beam of the laser is well suited for a number of surveying ap-plications—one laser, operating continuously, can replace two men and a tran-sit. Giant earth-boring machines and the holes they produce can be aligned through use of lasers. Bulldozers clearing land, graders leveling land, barges

71

and dredges working on dredging harbors or setting piers, pipe layers and ditch diggers are all making use of the laser as a simple method of alignment.

Another example is the Michaelson interferometer. This device is useful for measuring distances up to several hundred feet. Applications include: seismology where a stable source of coherent light can detect very small earth movements; metalworking which utilizes an interferometer to control the operations of a melting machine; flow rate control; and large scale movements when buildings or bridges sway.

Another application of lasers or a power source is in the drilling of dies for producing wire. The beam from a pulsed laser is focused on the die, usually a diamond, and repeated pulses are delivered until the dies are pierced. Prior to the use of this technique, fragile high speed drills were used. Still another application is the removal of flaws (piquer) in diamonds. The imperfections are removed by drilling a 0.002 in. hole to the site of the flaw with a laser and then removing it by one of several methods. The hole cannot be seen without magnification and the removal of the flaw enhances the value of the diamond with no loss in brilliance of the gem.

The laser is also attractive as a flexible welding tool which could replace electron beam welding. The disadvantages of electron beam welding include short working distances between source and weld and the required vacuum is great. A laser's ability to generate coherent light has given birth to the process of holography. Three-dimensional images are being used for display devices and as a method of spotting defects in automobile tires as well as in scientific research applications such as particle size measurement. A cube of crystal material can be used to record numerous holograms. The small size of the cube and the large number of three-dimensional images stored may herald an new era in information and data storage retrieval.

Lasers have also found use in the field of biology. For example, the retina of the eye is loosely attached to the choroid coat. The retina is of neurodermal origin while the choroid is ectodermal. In the embryo, these two join and subsequently throughout the life of the individual are held by a thin layer of connective tissue. In the adult, any number of circumstances, including trauma, can result in the separation of the retina from the body of the eye. This leads to loss of vision because the light cannot be properly focused upon the detached retina. In the past, retinas were reattached by using a long needle-like probe to weld the retina to the choroid with a scar. This worked quite well, producing one or more blind spots but allowing the proper focus to be attained once more. About 1950, the xenon photocoagulator was introduced, producing this same effect by means of a pulse of intense white light which, when focused by the lens on the retina, resulted in reattachment by coagulated blood in a fashion similar to a spot weld. More recently, retinal repair has been accomplished via the use of lasers as the light source. Ruby lasers were used first, then neodymium, and finally argon lasers. The real value of using the argon laser over the xenon photocoagulator is the size of the spot weld. An argon laser can produce welds much smaller than the size of a xenon weld, allowing finer "stitching" — this being of particular value around the fovea. In addition, neither anesthesia nor hospitalization is required with laser photocoagulation.

Skin cancers have also been experimentally treated with lasers. Since there is a difference between normal and cancerous skin cells, interest has been in the development of a dye or pigment that is completely selective for cancer cells. Partial results have been obtained and cancer cells can now be stained considerably darker than normal cells. The darker cells are more susceptible to the impact of a laser beam because they absorb more light energy and are more severely damaged than are normal unstained cells. Two types of cancer treatment have been practiced. A low energy beam has been used to selectively disrupt tumor cells. Higher energy beams are used to excise modules from deeper tissues.

Lasers have also been used in transillumination, which is a technique whereby a strong light is projected through soft tissue to aid in detecting tumors. The skin is relatively transparent to light, as is demonstrated by putting your thumb over a flashlight. Lasers hold promise for this type of examination, allowing, for example, an immediate examination for breast cancer without the potential hazard or wait associated with X-rays.

Some experimental work has already been applied in the field of dentistry. The glazing of teeth by a laser has been shown to reduce significantly the demineralization of enamel, and may also be effective against cavities. Dental cavities have been exposed to laser impact with favorable results. If the cavities can be retarded or stopped by laser impact, dentistry will have gained a valuable tool.

Laser technology is also applicable to cell identification. A new method of instant and positive identification of microorganisms and tissues is now being produced commercially. The sample is cooled to a low temperature and irradiated with ultraviolet laser light. Under those circumstances the sample itself produces phosphorescent light whose frequency and decay time are unique for the organism. A small computer matches the frequency and decay time data with information previously stored and can identify instantly the presence of specific microorganisms or tissues.

PRINCIPLES OF LASER OPERATION

The term <u>laser</u> is an acronym for "Light Amplification by Stimulated Emission of Radiation." It is a device which essentially generates and amplifies light.

Lasers vary greatly in output power, from a few milliwatts in the helium-neon gas laser, to thousands of watts in the carbon dioxide gas laser. They are capable of operating continuously or in pulses with millions of watts of power in each pulse.

The phenomenon of light comprises that portion of the electromagnetic spectrum which produces a visual effect. Quantum mechanics is the branch of science dealing with atomic and subatomic particles and describes the smallest indivisible quantity of radiant energy as one photon. The amount of energy, E, represented by one photon is determined by the frequency, ν, and Plank's constant, h:

$$E = h\nu \tag{5.1}$$

The frequency, ν, and wavelength, λ, of light are related by the velocity of light, c, so that when one is known, the other may be determined from the relationship:

$$c = \nu\lambda \tag{5.2}$$

Man has made use of almost the entire electromagnetic spectrum, from zero Hertz, such as direct current from storage batteries, to 10^{24} Hertz, the very hard X-rays used for nondestructive inspection of metal parts. Figure 5.1 reviews again the electromagnetic frequency spectrum and some of its uses and properties.

Light can be generated by atomic processes. It is these processes which are responsible for the generation of laser light. To understand how a laser operates one must examine the levels and then see how changes in these energy levels can lead to the production of laser light.

Assuming that the atom consists of a small, positively charged nucleus and one or more negatively charged electrons in motion about the nucleus, then the relationship between the electrons and the nucleus is described in terms of energy levels. Quantum mechanics predicts that these energy levels are discrete; they cannot have any arbitrary value. These energy levels are characteristic of specific atoms. A simplified energy level diagram for a single electron atom is illustrated in Figure 5.2.

The production of light by a quantum system is related to the energy levels of the system. Only frequencies of certain characteristic values can be generated. A relationship between these characteristic frequencies and the two energy levels of the atom was proposed by Niels Bohr:

$$\nu = \frac{E_2 - E_1}{h} \tag{5.3}$$

where ν = frequency of the photon, E_2 and E_1 = two allowed energies of the

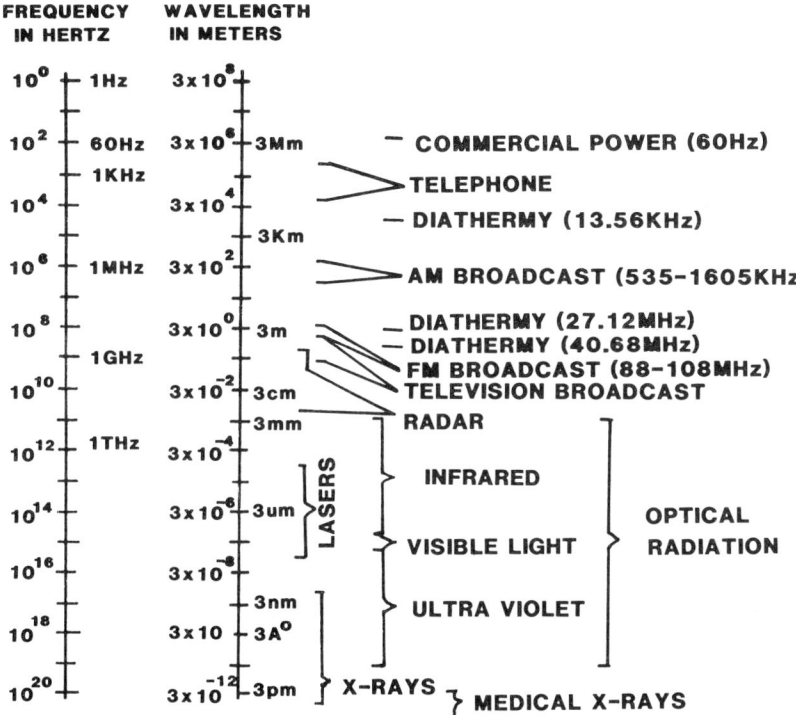

FIGURE 5.1. Electromagnetic spectrum and its benefits.

atom, and h = Planck's constant (h = 6.62 × 10^{-27} erg-sec). There are three distinct processes where atoms can change their energy state: absorption, spontaneous emission and stimulated emission. It is this third type of radiative transition that forms the basis for laser action. We shall briefly describe each form of transition.

With absorption, an electron can absorb energy from a variety of external sources. With a laser, two modes of supplying energy to the electrons are of prime importance. The first of these is the transfer of all of the energy of a photon to an orbital electron. The increase in the energy of the electron causes it to "jump" to a higher energy level; the atom is then said to be in an "excited" state. It is important to note that an electron accepts only the precise amount of energy that will move it from one allowable energy level to another. As such, only those photons of the energy or wavelength acceptable to the electron will be absorbed.

The second method used to excite electrons is an electrical discharge. Energy is supplied by collisions with electrons accelerated by the electric field. The result of either type of excitation is that through the absorption of energy,

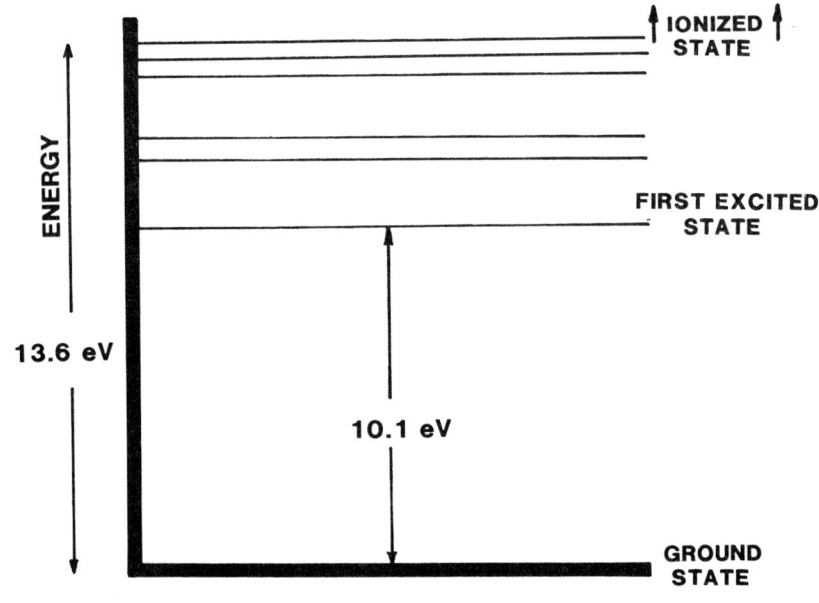

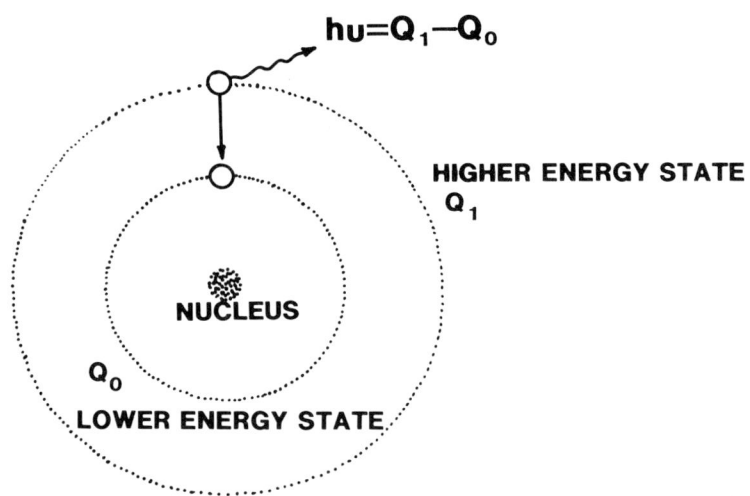

FIGURE 5.2. Energy level diagram of atom.

an electron has been placed in a higher energy level than that in which it had been residing, and hence, the atom is excited.

From thermodynamic principles, it is known that the entire atomic structure tends to exist in the lowest energy state possible. An excited electron in a higher energy level will thus attempt to "de-excite" itself by any of several means. A portion of the energy may be converted to heat. Another means of de-excitation is the spontaneous emission of a photon. The photon released by an atom as it is de-excited will have a total energy exactly equal to the difference in energy between the excited and lower-energy levels. This release of a photon is referred to as <u>spontaneous emission</u>. An example is the neon sign. Atoms of neon are excited by an electrical discharge through the tube. They de-excite themselves by the emission of photons of visible light. Note that the exciting force is not of a unique energy, so that the electrons may be excited to any one of several energy levels. The photons released in de-excitation may have any of these several discrete frequencies. If enough discrete frequencies are present in the appropriate distribution, the emission may appear to the eye as "white" light.

A photon released from an excited atom could, upon interacting with a second, similarly <u>excited</u> atom, trigger the second atom into de-exciting itself with the release of a photon. The photon released by the second atom would be identical in frequency, energy, direction, and phase with the triggering photon, and the triggering photon would continue on its way, unchanged. In other words, where there was one, now there are two. Figure 5.3 illustrates this concept. These two photons could then proceed to trigger more atoms through stimulated emission.

If an appropriate medium contains a great many excited atoms and de-excitation occurs *only* by spontaneous emission, the light output will be random and approximately equal in all directions. The process of stimulated emission, however, can cause an amplification of the number of photons traveling in a particular direction, known as a photon cascade. A preferential direction is established by placing mirrors at the ends of an optical cavity. Photons not normal (perpendicular) to the mirrors will escape. Thus, the number of photons traveling along the axis of the two mirrors increases greatly and light amplification by the stimulated emission of radiation occurs.

The process of stimulated emission does not produce very efficient light am-

FIGURE 5.3. Illustrates photon multiplication via stimulated emission.

plification unless "population inversion" occurs. If only two of several million atoms are in an excited state, the chances of stimulated emission occurring are infinitely small. The greater the fraction of the population of atoms in an excited state, the greater the probability of stimulated emission. In the normal state of matter, the population of electrons will be such that most of the electrons reside in the ground or lowest energy levels, leaving the upper levels somewhat depopulated. When electrons are excited and fill these upper levels to the extent that there are more atoms excited than not excited, the population is said to be inverted.

It is important to note that the light output of a laser is considerably different from the output of ordinary light sources. Four properties characterize the laser's output: small divergence, monochromaticity, coherence, and high intensity. When light emerges from the laser, it does not diverge (spread) very much. Consequently, the energy is not greatly dissipated as the beam travels. Laser beam divergence is measured in units of milliradians. There are 2π radians in a circle so one milliradian equals about 3 minutes of arc. A typical He-Ne laser has a rated divergence of 0.5–1.5 milliradians.

Laser light is very close to being monochromatic, i.e., being of one color, or wavelength. Actually, very few lasers produce only one wavelength of light. A typical He-Ne laser emits light at 632.8 nm, which is orange-red, and at 1,150 nm and 3,390 nm in the near and middle infrared regions. The He-Ne laser is usually designed to emit ony one of the three wavelengths of light and the variation in this wavelength is slight.

Coherence refers to the relationships between two wave forms. Two waves with the same frequency, phase, amplitude, and direction are termed spatially coherent. No source of perfectly spatially coherent light is yet known; laser light approximates this and for practical purposes, it can be considered perfectly coherent.

Laser light can be very intense. The sun emits about 7×10^3 W/cm²/Sr/μm at its surface. Lasers are presently capable of producing more than 1×10^{10} W/cm²/Sr/μm. The magnitude 1×10^{10} W/cm²/Sr/μm represents only a single pulse of light. Energy is a measure of capacity for doing work and is categorized as potential or kinetic energy. It is commonly measured in joules (J). Power is the rate at which work is being done and is measured in watts (W). The following relationships hold:

1 joule = 1 watt-second
1 watt = 1 joule-second

A laser capable of emitting 10 joules in one second can be termed a 10-watt laser. If those same 10 joules are emitted as a single pulse of 1/100th second duration, then the laser can be termed a 1,000-watt laser. The output of pulsed lasers is usually indicated in terms of J/cm². The effect of the laser pulse strongly depends on the amount of time it takes to deliver the pulse. Consequently, pulsed laser output is sometimes referred to in terms of $(J/cm^2)/sec$ or W/cm^2.

Any laser has three basic components: (1) an active lasing medium; (2) an input energy source (referred to as a pumping system); (3) an optical cavity. Lenses, mirrors, shutters and other accessories may be added to the system to obtain more power, shorter pulses, or special beam shapes.

A lasing medium, to have suitable energy levels, must have at least one excited state (metastable state) where electrons can be trapped and do not immediately and spontaneously drop to lower states. Electrons may remain in these metastable states from a few microseconds to several milliseconds. When the medium is exposed to the appropriate pumping energy, the excited electrons are trapped in these metastable states long enough for a population inversion to occur, i.e., there are more electrons in the excited state than in the lower state to which these electrons decay when stimulated emission occurs. Figure 5.4 illustrates a three-level energy diagram for a laser material. This is just one of the many possible systems of energy levels. Although laser action is possible with only two energy levels, most such actions involve four or more levels. There are four common types of lasers which are classified according to the state of their lasing media. The first of these are solid state lasers which employ a lasing material distributed in a solid matrix. One example is the ruby laser, using a precise amount of chromium impurity distributed uniformly in a rod of crystalline aluminum oxide. The output of the ruby is primarily at a wavelength of 694.3 nm, which is deep red in color (refer to Figure 5.5).

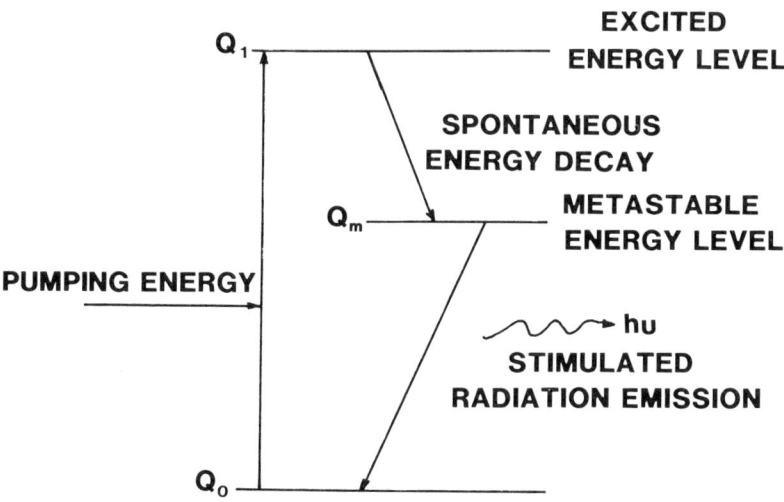

FIGURE 5.4. Three-level energy diagram for laser action.

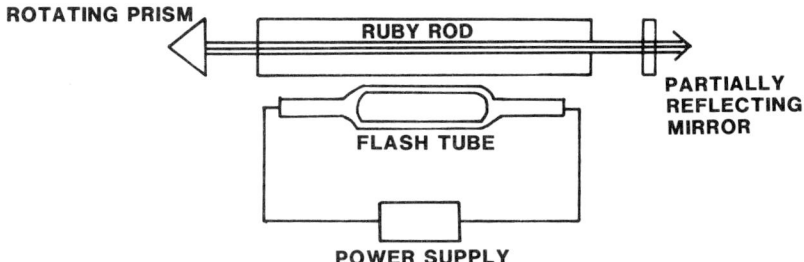

FIGURE 5.5. Illustrates solid state laser with optical pumping.

The second class are gas lasers which use a gas or a mixture of gases within a glass tube. Common gas lasers include the He-Ne laser, with a primary output of 632.8 nm and the CO_2 laser, which radiates at 10,600 nm, in the infrared (refer to Figure 5.6). Argon and krypton lasers have outputs in the blue and green regions.

Semiconductor lasers are devices that consist of two layers of semiconductor material sandwiched together. One material consists of an element with a surplus of electrons, the other with an electron deficit. Two outstanding characteristics of the semiconductor laser are its high efficiency and small size. Typical semiconductor lasers produce light in the red and infrared regions.

Liquid lasers are relatively new, and the lasing medium is usually a complex organic dye. The most striking feature of the liquid lasers is their "tunability." Proper choice of the dye and its concentration allows light production at almost any wavelength in or near the visible spectrum.

The pumping system is used to raise electrons to a higher energy level. These systems pump energy into the laser material, increasing the number of electrons trapped in the metastable energy level. When the number of electrons in the metastable energy exceeds those in the lower level, a population inversion exists, and the laser operates. Laser action can occur only when a population inversion has been established in the lasing medium. This population inversion can be established by pumping energy into the lasing medium. Several methods of pumping are commonly used. Optical pumping is used in solid and liquid lasers. A bright source of light is focused on the lasing medium. Those incident photons of correct energy are absorbed by the electrons of the lasing material and cause the latter to jump to a higher level. Xenon flashtubes similar to strobe lights used in photography, but more powerful, are commonly used as optical pumps for solid state lasers. Liquid lasers are usually pumped by a beam from a solid state laser. Electron collision pumping is utilized in gas lasers. An electrical discharge is sent through the gas-filled tube. The electrons of the discharge lose energy through collisions with gas atoms or molecules and the atoms or molecules that receive energy are excited. Electron collision pumping can be done continuously and can therefore lead to a continuous laser output.

Chemical pumping is based on energy released in the making and breaking of chemical bonds (i.e., Hydrogen-Fluoride lasers).

Once the lasing medium has been pumped and a population inversion established, lasing action may begin. If, however, no control were placed over the direction of beam propagation, photon beams would be produced in all directions. This is referred to as <u>superradiant lasing</u>. The direction of beam propagation can be controlled by placing the lasing medium in an optical cavity formed by two reflectors facing each other along a central axis.

Photon beams which are produced along the cavity axis are reflected 180° at each reflection and travel once more through the lasing medium causing more stimulated emission. Thus, the beam grows in magnitude with each traverse of the lasing medium.

Since the reflectors are not 100% reflective, some photons are lost by trans-

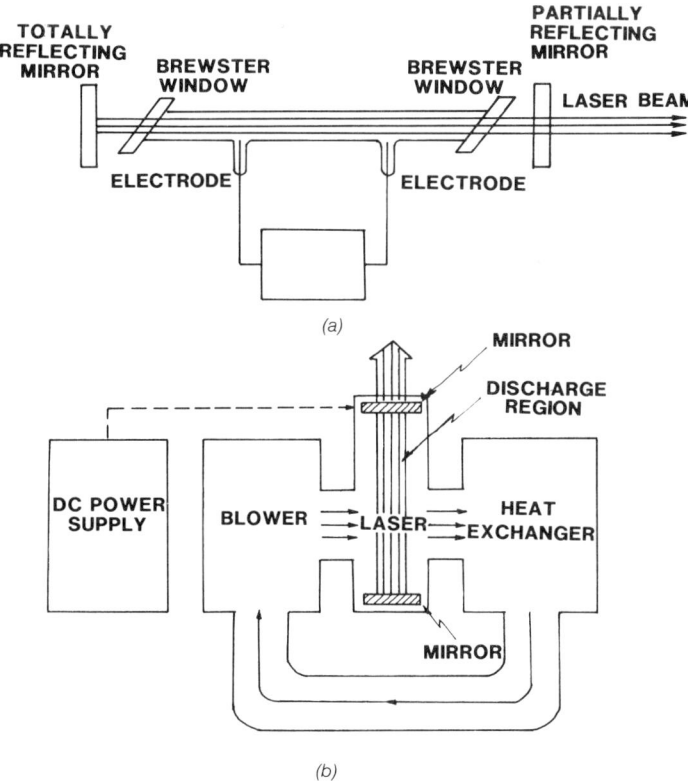

FIGURE 5.6. (a) He-Ne laser with electron collision pumping; (b) CO_2 gas transport laser.

mission through the mirrors with each passage. If the pumping is continuous, an equilibrium stage will be reached between the number of photons produced by atoms raised to the excited state and the number of photons emitted and lost. This results in a continuous laser output and is usually used only with low power input levels. Higher power inputs usually are achieved in the form of a pulse, and the output is also in pulse form. One of the mirrors in the system is usually made more transparent than the other, and the output, pulsed or continuous, is obtained through this reflector.

The quality of the optical cavity can be altered by placing a shutter between the mirrors. This enables the beam to be turned off and on rapidly and creates pulses with a duration of tens of nanoseconds to tens of monoseconds. This mode of operation is called Q-switched. This technique is used to produce an exceptionally high-power output pulse. The term Q as applied to lasers is derived from the more familiar Q of electrical circuits. Lasers are resonant cavities and in a similar way, many electrical devices are resonant. The Q is a numerical index of the ability of the resonant cavity to store energy at the output frequency. The higher the Q, the more effective the power concentration at the resonant frequency. Q-switching in lasers refers to the method of laser operation in which the power of the laser is concentrated into a short burst of coherent radiation. A Q-switch is a device which interrupts the optical cavity for a short period of time during pumping.

Reflectors may consist of plane mirrors, curved mirrors, or prisms. The mirror coating may be of silver, if laser output power is low, but higher powers may require dichroic material. A dichroic material is a crystalline substance in which two preferred states of polarization of light may be propagated with different velocities and, more important, with different absorption. By appropriate choice of material and thickness, the light impinging upon the dichroic coating may be either totally absorbed or totally reflected. The first ruby lasers were constructed with the crystal ends polished optically flat and silvered. Semiconductor lasers use a similar technique. Gas lasers may use mirrors as seals for the ends of the gas tube or may utilize exterior mirrors. Further discussions are given by Cheremisinoff (1986).

PRINCIPLES OF COHERENCE, INTERFERENCE AND LASER OPTICS

As noted in earlier discussions, we may treat light as a wave phenomenon. The properties and mathematical laws defining waves can therefore be applied to characterizing light. Consider an ideal elementary sinusoidal wave form. The characteristics of this wave are its amplitude A (its height above the zero line) and its wavelength λ (the distance from wave peak to wave peak). In this case, the amplitude varies with time in a constant sinusoidal manner and with the maximum remaining the same. The wavelength also remains constant, as the wave propagates. Figure 5.7(A) illustrates these parameters.

The term <u>coherence</u> refers to the interaction between two or more waves. Consider two waves, one superimposed over the other on the same line of propagation. Figure 5.7(B) shows the two waves with the superimposition modified for clarity. For our discussions, consider points *only* with respect to the direction of propagation. During a time, each of the waves advances along

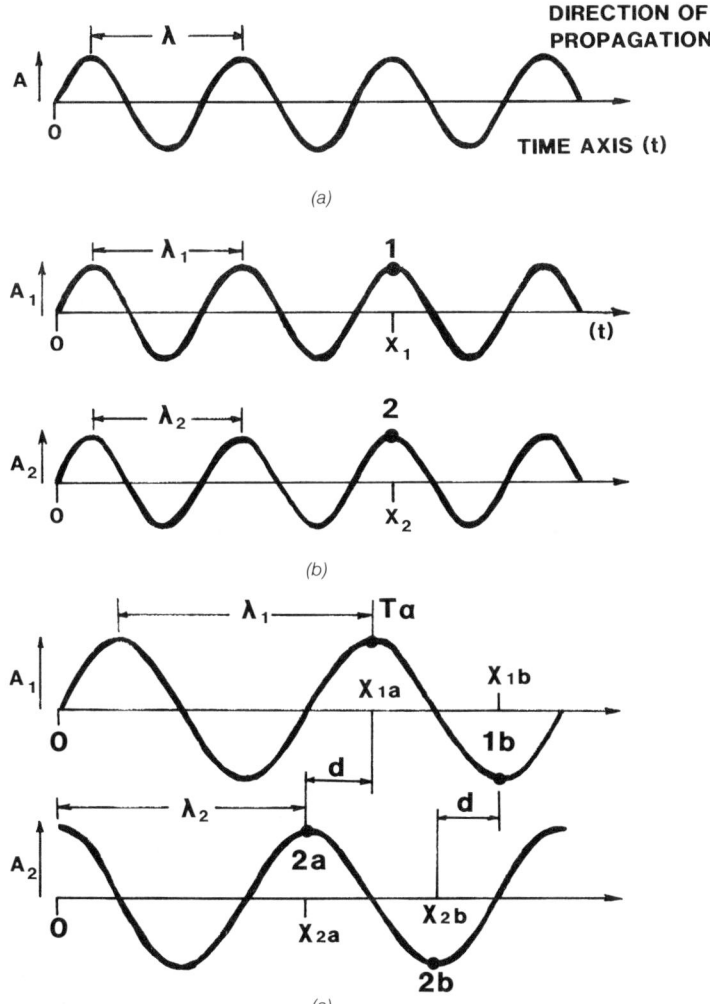

FIGURE 5.7. (a) shows a sine wave; (b) illustrates temporal coherence, no phase difference; (c) temporal coherence, constant phase difference.

the line of propagation to a wave peak, point 1 on wave A_1 and point 2 on wave A_2. Both waves travel at the same speed and have the same wavelength and, if allowed to continue, the wave peaks will occur at the same places all the way down the line. These waves are in phase and are coherent. This type of coherence, with respect to time, is referred to as temporal coherence.

Two or more waves are said to be coherent if the "phase difference" between two pairs of points, one on each wave, remains constant. In this example, where the waves are in phase, the phase difference is zero throughout and therefore constant.

Figure 5.7(C) shows a case where one wave "lags" the other slightly, but the waves are still temporally coherent because the sets of points remain at a constant distance (d) apart along the line of propagation. They are not in phase but have a constant phase difference.

We now examine two waves superimposed on each other along the same line of propagation. As a wave travels through a medium, it propagates a disturbance along the direction of propagation, with the amount of disturbance depending on the amplitude of the wave. When the amplitude is positive (i.e., when a peak is formed or the wave point is moving up), it is termed positive disturbance. Alternately, when the amplitude is negative, it is termed negative disturbance.

Superimposing the two waves on one another along the same line of propagation results in the addition of their amplitudes. If there are two positive amplitudes, they total to a greater positive amplitude. Two negative amplitudes follow the same formula, adding to a greater negative amplitude. If a negative amplitude is paired with a positive, the difference between the two is found for the total amplitude. This interaction is known as interference. Constructive interference results from adding two wave amplitudes of the same sign ($+$ or $-$). Destructive interference results from adding two amplitudes of different signs.

Now consider two waves traveling along two parallel lines of propagation, with equal amplitudes, wavelengths, and in phase (as shown in Figure 5.8). Following the points A, B and C, the distance between them remains constant. The points are not only in phase along the path of propagation, but also in a direction perpendicular to the paths of propagation. Since these points are in phase perpendicular to the path of propagation, they are laterally coherent.

Temporal coherence remains the same along the direction of propagation as does lateral coherence, but now lateral coherence is allowed to relate to any direction perpendicular to the line of propagation. The result of combining these two concepts, lateral and temporal coherence, produces what is called spatial coherence in which two or more points are in phase, a) along the direction of propagation (temporal), and b) along *any* direction perpendicular to the direction of propagation (lateral). This results in a three-dimensional depiction of light as illustrated in Figure 5.9.

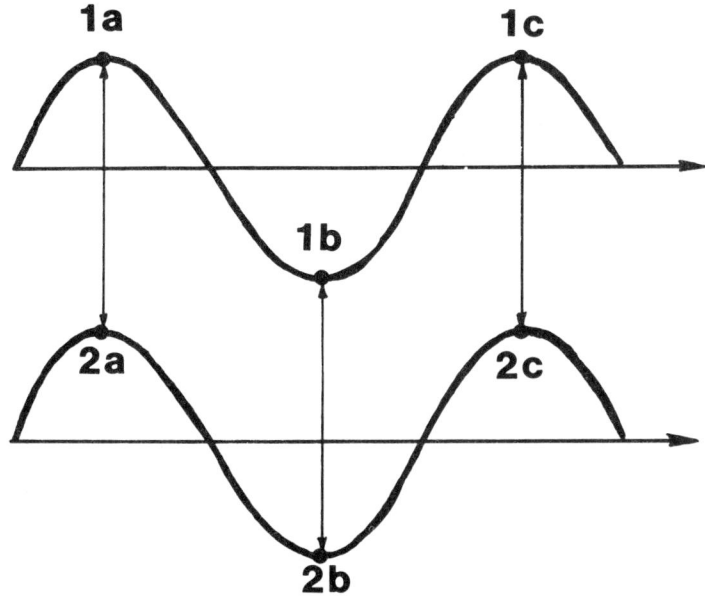

FIGURE 5.8. Illustrates lateral coherence.

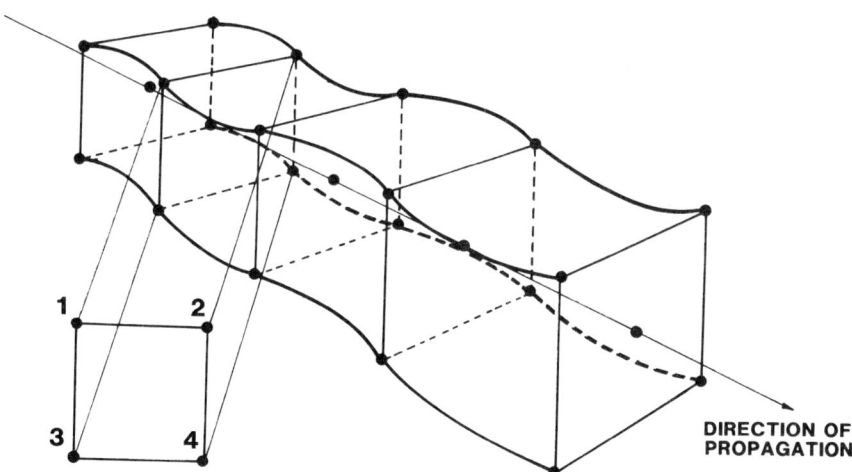

DIRECTION OF
PROPAGATION

FIGURE 5.9. Illustrates spatial coherence.

It is important not to lose sight of the fact that all waves have a definite band-width B. This bandwidth determines the phase change that the wave can take and thus provides a measure of the wave's coherency. We can approximate the time during which a wave remains reasonably coherent from the reciprocal of the bandwidth, $T_c \cong 1/B$. The length or distance over which a wave would remain coherent, then, would be the coherence time T_c multiplied by the velocity of the wave. With light, the velocity is $c = 2.99774 \times 10^8$ m/sec. The coherence length L_c would then be equal to $L_c = v/B = c/B$, where v is the velocity of the wave and c the velocity of light.

The number of wavelengths n over which the wave is coherent is

$$n = c/B\lambda \tag{5.4}$$

Ordinary light is not very coherent and its uses are limited by two physical rules, namely, Abbe's sine condition and the second law of thermodynamics. Both of these were formulated well before coherent light was understood. These conditions define both the ability to focus ordinary light and the amount of energy that can be transferred from a source of ordinary light to some material at another point in space.

Coherent light, such as that generated by a laser, is not subject to such restrictions. It can be focused into an extremely narrow beam, and onto a spot limited only by diffraction effects. The second law of thermodynamics states that when temperatures of the source and sink are equal, no further transfer of energy can occur. This does not apply strictly to laser light; the temperature at the focused image of the laser can be made as high as the diffraction limitation allows. Because of diffraction limited focusing, laser light can be focused down to a minimum cross section of about one square wavelength. This opens the possibility of attaining high density levels over extremely small areas.

We now briefly describe the interaction of optical radiation with matter. First, the laws of reflection must be stated. One of the basic laws is that the angle of incidence is equal to the angle of reflection. A second law states that incident and reflected rays, as well as those normal to the reflecting surface, all lie in the same plane (called the plane of incidence). These laws are known as Huygen's principle.

Light travels in straight lines and is reflected at the interface between two media according to these laws. When it enters a different medium across an interface, its velocity changes. This phenomena results in the refraction or bending of the light wave front as it passes obliquely from one medium into another. The ratio of the velocity of light in one medium to its velocity in a second medium is the index of refraction.

The velocity of light in a given medium also depends on its wavelength. This dependence is related to its dispersion and is accounted for by saying the index of refraction varies with wavelength. In geometrical optics, all the properties of lenses and mirrors can be explained by knowing that light travels in straight lines and obeys the laws of reflection and refraction when diffraction and interference effects are neglected.

When light traveling through one substance obliquely enters a second substance having a higher index of refraction, it is bent toward the perpendicular, i.e., toward the normal to the surface. When it enters a substance having a lower index of refraction, it is bent away from the perpendicular. Thus, when a light beam passes obliquely from water or glass into air, the refracted ray is bent away from the perpendicular. The relationship between the angles and indices of refraction is given by Snell's law:

$$\eta_1 \sin \theta_i = \eta_2 \sin \theta_+ \tag{5.5}$$

where

η_1 = index of refraction of medium 1 with respect to a vacuum
η_2 = index of refraction of medium 2 with respect to a vacuum
θ_i = angle of incidence
θ_+ = angle of refraction

As the angle of incidence θ_i increases, the angle of refraction θ_+ increases to a value where the refracted beam just grazes the surface of the interface between the two materials. The angle of incidence θ_i which produces an angle of refraction θ_+ of 90° is the critical angle. For a water/air interface, the critical angle is about 49 degrees. If this critical angle is exceeded, the beam does not leave the material at all but is instead totally reflected internally.

The index of refraction of air with respect to a vacuum is unity, and the index of refraction of water with respect to a vacuum is 1.33. The relative index of refraction of air with respect to water (i.e., for a light ray going from water into air) is about 0.75. The critical angle for a water-air interface is therefore 48.6°, which the reader can verify from Equation (5.5).

Diffraction refers to phenomena in which light or other electromagnetic radiation is bent around an obstacle instead of exhibiting a simple straight-line propagation predicted by geometrical optics.

If light passes near the edge of an object, light and dark bands are seen in the region of the geometric shadow. The light can thus be "bent" around an opaque object. The light is bent (or diffracted) by obstacles, in a fashion similar to the way waves on water are bent around a pier in their path.

From Huygen's principle, when light emerges from an aperture or the edge of a barrier, each point along a plane perpendicular to the direction of propagation of the incident light may be regarded as a new source. The amplitude of the radiation from these new sources arriving at the viewing screen depends on the distance from the new source to the point at which they strike the screen. The sum of the interference of the light waves from a slit or barrier edge results in an illumination pattern of maximum and minimum intensity. In the case of a slit, the spacing between regions of maximum and minimum intensity is inversely proportional to the width of the slit. A diffraction pattern of light from two slits is simply the interference pattern from the two slits superimposed on the diffraction patterns from the individual slits. An elaboration of the double-slit includes the use of a number of slits equally spaced. Such an array results in a diffraction grating, and the diffraction pattern obtained is the result of multiple interference of a large number of slits so that the maximum and minimum are much sharper than before. Discussions of diffraction and laser optics are given in the references noted at the end of this chapter.

HAZARDOUS EFFECTS FROM LASER RADIATION

Both laser radiation and optical radiation generated by ultraviolet, infrared, and visible optical sources have similar physiological effects. The unique biological implications attributed to laser radiation are generally those resulting from the very high beam collimation, beam intensities and monochromaticity of many lasers. Such sources differ from conventional sources of optical radiation primarily in their ability to attain highly coherent radiation. The increased directional intensity of the optical radiation generated by a laser results in concentrated optical beam irradiances at considerable distances.

The exact biologic effects from laser radiation are still undefined. It should be noted, however, that in the medical field, laser energy is being used to irradiate individual tumors, to supplement X-ray treatment of various tumors, to treat some types of eye conditions, such as vascular lesions and detached retinas, and as therapy for certain types of chronic skin lesions.

Laser radiation is in fact different from ionizing radiation (such as X- and gamma rays), although very high irradiances can produce ionization in air and other materials. The biologic effects of laser radiation are essentially those of visible, ultraviolet or infrared radiation upon tissues. In contrast, however, radiant intensities typically produced by lasers are of magnitudes that could previously be approached only by the sun, nuclear weapons, burning magnesium, or arc lights. This fact makes lasers potentially hazardous. Laser radiation incident upon biologic tissue will be reflected, transmitted, and/or absorbed. The degree to which each of these occurs depends upon various properties of the tissue. Absorption is selective, as in the case of visible light; darker pigmented tissue absorbs more energy, for example.

Adverse thermal effects resulting from exposure of the skin to radiation from 315 nm to 1 mm may vary from mild reddening (erythema) to blistering and charring. The degree of damage depends on the exposure dose rate and the dose (amount of energy) transferred and conduction of heat away from the absorption site. Adverse skin effects resulting from exposure to actinic ultraviolet radiation (200–315 nm) vary from erythema to blistering depending upon the wavelength and total exposure dose.

The organ that is most vulnerable to laser radiation is the eye. Most higher energy X-rays and gamma rays pass completely through the eye, with relatively little absorption. Absorption of short-ultraviolet (UV-B and UV-C) and far-infrared (IR-B and IR-C) radiation occurs principally at the cornea. Near-ultraviolet (UV-A) radiation is primarily absorbed in the lens. Light is refracted at the cornea and lens, and absorbed at the retina; near-infrared (IR-A) radiation is also refracted and is absorbed in the ocular media and at the retina.

Retinal damage from lasers is generally believed to be limited largely to the spectral regions of 400–700 nm for light and 700–1400 nm for laser-infrared radiation. The effect upon the retina may be a temporary reaction without residual pathologic changes, or it may be more severe with permanent pathologic changes resulting in a permanent scotoma. The mildest observable reaction may be simple reddening. If the retinal irradiance continues, lesions may occur which progress in severity from edema to charring, with hemorrhage and additional tissue reaction around the lesion. Excessive radiant exposures can cause gases to form near the site of absorption which may disrupt the retina and may alter the physical structure of the eye. Portions of the eye other than the retina may be selectively injured, depending upon the region where the greatest absorption of the specific wavelength of the laser energy occurs and the relative sensitivity of tissue affected.

With ultraviolet radiation (200–400 nm), symptoms similar to those observed in arc welders can occur. It may cause severe acute inflammation of the eye and conjunctiva. UV-B and UV-C radiation does not reach the retina. Near-ultraviolet radiation (UV-A) is absorbed primarily in the lens, which causes the lens to fluoresce. Very high doses can cause corneal and lenticular opacities.

Far-infrared radiation, IR-B and IR-C, covers the region of 1.4–1000 μm. Absorption of far-infrared radiation produces heat, with its characteristic effect on the cornea and the lens of the eye. The 10.6 micrometer wavelength from the carbon-dioxide laser is absorbed by the cornea and conjunctiva and may cause severe pain and destructive effects.

The ocular media (OM) of the human eye transmits wavelengths over the range of 400 to 1,400 μm. The transmission exceeds 90% in the range of 500 to 950 μm. There are two peaks in the near infrared (IR), a large one at 1,100 μm, and a smaller one at 1,300μm. The transmission of the OM is similar to that of physiological saline. Wavelengths below 400 μm are absorbed by the cornea, which is susceptible to damage by ultraviolet (UV). Inflammation, known as keratitis, is caused by overexposure of the cornea to UV light. At wavelengths above 1,400 μm, in the near- and far-IR, light is not focused on the retina, and, hence, does not constitute a risk to retinal tissue. However, radiation in the near-IR can results in opacification of the lens, a hazard referred to as infrared cataract, also sometimes referred to as glassblower's cataract. Infrared cataract is generally associated with chronic exposure to IR. The iris absorbs the IR and transmits the heat energy to the lens, which becomes overheated. If the lens remains at temperatures above normal for extended periods, heat denaturation of the lens proteins occurs. Hence, long-term exposure to near-IR should be avoided. Microwaves also cause opacification of the crystalline lens. The lens of the human eye is sensitive to opacification from three regions of the electromagnetic spectrum, namely, X- and gamma rays (radiation cataract), infrared cataract, and microwave cataract.

The normal vision for an unaccommodated human eye can be simulated by a positive lens of approximately 17 mm in focal length. The amount of radiant energy entering the eye is controlled by the iris, whose contraction and dilation constitutes the pupillary reflex. Pupillary diameters range from about 2 to 8 mm, depending upon the brightness of the environment. For example, in bright sunlight the pupil may contract to about 2 mm, whereas, after prolonged absence of light, the pupil dilates to 8 mm or more. The status of the pupillary diameter at the time of exposure is critical, since the area is proportional to the square of the diameter. The net result is that an 8 mm pupil will admit 16 times more light than a 2 mm pupil.

There are basically three types of exposures encountered from laser beams, depending upon the mode of operation. These are: Q-switching (exposure time less than 100 nanosecond), normal multiple spike mode (exposure time ranging from 100 μsec to 1 or 2 millisecond), and the continuous-wave (C-W) mode, where the time of exposure may vary from a few msec to seconds or minutes, depending upon conditions. The degree of injury to the retina is a function of the physical parameters of power density, exposure time, wavelength, and image diameter on the retina. Extent of injury also depends on whether an exposed individual is lightly or heavily pigmented in the retina and choroid; whether he has a large or small pupillary diameter; and whether his exposure is limited by the blink reflex (usually taken as about 150 msec in man) and by other physical parameters.

The type and extent of damage inflicted to the eye ranges from a small and inconsequential retinal burn in the periphery of the fundus, to severe damage of the macular area, with consequent loss of visual acuity, up to massive hemorrhaging and extrusion of tissue into the vitreous, with possible loss of the entire eye. The Q-switched laser, because of its high power density and short exposure time, represents potentially the greatest hazard.

The Armed Forces have adopted safety criteria for the protection of military personnel, and the American Conference of Governmental Industrial Hygienists has issued a guide for codes or regulations on laser safety; also, many governmental and industrial concerns have issued their own manuals on laser safety. The United States of American Standards Institute (USASI) also has a committee to study laser safety and make recommendations. The Executive Office of the President, Office of Telecommunications Management, has established an Electromagnetic Radiation Management Advisory Council (ERMAC) to advise on the hazards of electromagnetic radiation. The National Council on Radiation Protection and Measurements has also considered expanding its interests to include lasers. The Radiation Control for Health and Safety Act of 1968, Pulic Law 90–602, has been enacted to protect the public from the dangers of electronic product radiation. Various parameters governing laser safety standards include:

1. Power density (watts per square centimeter) at the cornea or energy in joules (J) entering the eye
2. Exposure time
3. Wavelength or spectral distribution at cornea
4. Transmission through ocular media (OM) as function of wavelength
5. Diameter of pupil
6. Diameter of retinal image
7. Absorption or transmission by retinal pigment epithelium as a function of wavelength
8. Absorptance or transmission by choroid as a function of wavelength

Threshold damage to the retina is defined as barely ophthalmoscopically visible damage at a selected time after exposure. The problem then is to determine quantitatively the energy density at the retina in joules per square centimeter needed to produce threshold damage in a given exposure time. This data can be used to calculate the power density at the cornea or the total energy entering the eye for a given diameter of the pupil. Thus, standards may be defined for daytime and nighttime exposures (diameter of the pupil ranging from 2 to 8 mm), wavelength or spectral distribution, mode of operation (Q-switched, multiple spike, pulse train, or C-W), and optical factors in the near or far field which determine image size on the retina. For C-W lasers, where exposure times may be relatively long (a few milliseconds to many seconds or even minutes), the diameter of the image on the retina is extremely important. For a given power density on the retina in watts per square centimeter, the time taken to reach temperature equilibrium and the magnitude of the equilibrium temperature attained are both strong functions of the image size. For example, a 10μ or 20μ diameter spot size on the retina will come to temperature equilibrium in milliseconds, whereas for the same retinal input in watts per square centimeter, a $1,000\mu$ (1-mm diameter) spot size requires 7 to 10 seconds to attain equilibrium. Moreover, the temperature rise will be a hundredfold less for the 10μ diameter than for the $1,000\mu$ diameter, again assuming the same power density on the retina. Optical instrumentation, such as binoculars and telescopes, increases the image size of a radiant source on the retina by a large factor, even though the energy density on the retina may remain constant. For point sources not resolvable by the eye, the power density on the retina will be increased by the ratio of the collecting power of the instrument to that of the eye alone.

The Federal government's safety performance standard is incorporated in the Bureau of Radiological Health (BRH) laser product standard. It requires manufacturers to market laser systems which have "accessible emissions" which are only sufficient to accomplish the task and which have secondary (i.e., reflected) beams that do not expose personnel to levels above the exposure limits. This standard and the proposed OSHA Standard for laser safety in the workplace make use of a hazard classification scheme adapted from the earlier ANSI Standard Z-136.1 (1973). The fundamental concept is that safety restrictions must become more stringent for higher-power, higher-risk laser devices.

Although ANSI began with five hazard classes, only four are now used in the Federal standards. Class I Laser Products are basically safe; even long-term viewing of emissions from enclosed higher power systems are not considered hazardous. Class II Laser Products are basically HeNe or other visible CW with an output power not exceeding 1 mW. Class III (medium power) Laser Products present a hazard only for intrabeam viewing, i.e., CW lasers with output powers below 500 mW. Class IV (high power) Laser Products may present any one or all of the following three dangers: hazardous diffuse reflections, substantial skin hazards, or fire hazards. Practically all material processing lasers are considered Class IV lasers unless they are enclosed. If the enclosure is sufficiently interlocked to assure no hazardous accessible emissions, then the system could be classified as a Class I product. The BRH and OSHA require certain warning labels on the Class II, III, and IV lasers, but if these are enclosed to Class I specifications, then the warning label need only be on an inside access panel to warn maintenance personnel if they should attempt to gain operational access to the laser.

It is generally recommended to use glass or plastic filter windows to view the laser target area. Clear plastics are commonly used to protect against reflected levels of CO_2 (10.6 μm) laser radiation. Some glass filter materials that are useful with ruby and neodymium lasers are Schott KG-3 for 1064-nm filtration (1.5 OD per mm of thickness), Schott BG-18 for 694.3 nm (2.0 OD per mm of thickness at 694.3 nm and greater attenuation at 1064 nm) and American Optical OLF-61 for 1064 nm (1.5 OD per mm of thickness). Figure 5.10 gives the spectral transmittance curves for these filters relative to the phototropic response of the eye to illustrate their total visible transmission.

An additional note to consider is that many of the materials required for chemical laser generation, as well as products produced by these lasers, are potentially hazardous to operating personnel. Exposures resulting from contact with laser fuels or exhaust products can include: carbon monoxide, meth-

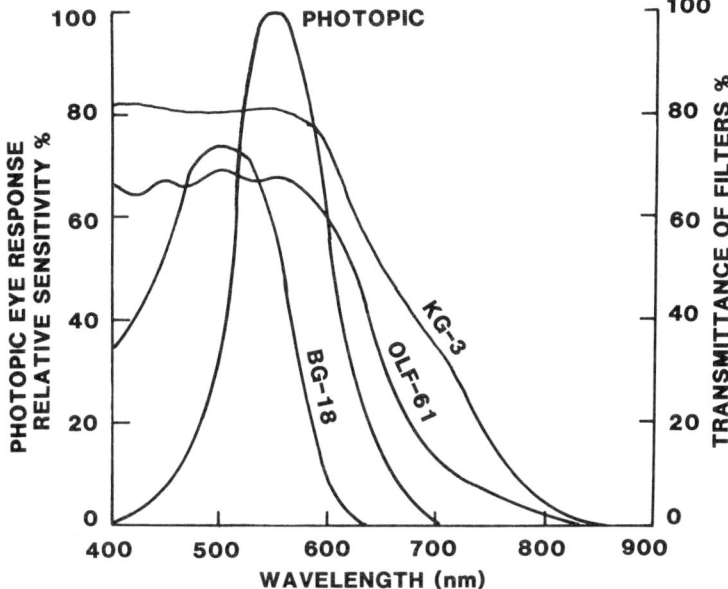

FIGURE 5.10. Spectral transmittances of various types of glass protective filters. Transmittances are compared against visual sensitivity of the eye.

ane, sulfur hexafluoride, nitrogen, helium, fluorine, lithium, carbon disulfide, hydrogen, carbon dioxide, fluorides, nitrogen oxides, sulfur compounds and refrigerants. Finally, beam interaction with metallic targets can produce airborne contaminants similar to those generated as the result of arc-welding on similar metals.

The allowable time-weighted exposure concentrations for various byproducts of laser fuels and beam generation are listed in Table 5.1. (Refer to OSHA Standards, Code of Federal Regulations, Subpart G, Section 1910.93.) Most of these values are time-weighted concentrations for an 8-hour work day, 40-hour work week and that occasional excursions above these values are permitted. Although general guidelines have been established to limit the magnitude of these excursions, definitive information is available only for a limited number of materials. In addition, for some contaminants no excursion is allowed. Those exposure values annotated with an asterisk in Table 5.1 have been given ceiling designations. This means that time-weighting is not applicable and that all exposure levels should fluctuate below the designated value.

Exposure of operating personnel to hazardous materials associated with laser operation should be kept below established safe concentrations. This can be accomplished by local or dilution ventilation, isolation, shielding, the use of personal protective devices, and other engineering controls. A frequently

Table 5.1. Various contaminants associated with laser production.

Contaminants	Probable Source	Allowable Time-Weighted Exposure
Asbestos	Target backstop	5 fibers/cc*
Beryllium	Firebrick target	0.002 mg/m³
Cadmium oxide fume	Metal target	0.1 mg/m³
Carbon monoxide	Laser gas	50 ppm
Carbon dioxide	Active laser medium	5.000 ppm
Chromium	Metal targets	0.5 mg/m³
Cobalt, metal fume, and dust	Metal targets	0.1 mg/m³
Copper fume	Metal targets	0.1 mg/m³
Fluorine	Of high-frequency chemical laser	0.1 ppm
Hydrogen fluoride	Active medium of laser	3 ppm
Iron oxide fume	Metal targets	10 mg/m³
Manganese	Metal targets	5 mg/m³**
Nickel	Metal targets	1 mg/m³**
Nitrogen dioxide	GDL discharge	5 ppm
Ozone	Target and Marx generators	0.1 ppm
Sulfur hexafluoride	Saturable absorber	1.000 ppm
Uranium (soluble/insoluble)	Target	0.05/0.25 mg/m³
Vanadium fume (dust/fume)	Target	0.5/0.1 mg/m³**
Zinc oxide fume	Target	5 mg/m³

* > 5 μ meters in length.
** Denotes ceiling concentration.

used method of control is dilution ventilation. For example, over-exposure to airborne contaminants from lased target materials located outdoors would not be expected because of both dilution ventilation and isolation. For indoor operations, local exhaust ventilation is one of the most effective control methods. Airborne contaminants are captured near the point of evolution before they have a chance to be distributed throughout the room. Mechanical dilution (fan) ventilation is generally not an acceptable method for indoor control of the more toxic materials (exposure limit of 100 ppm and below, as a rule of thumb) or where contaminants are generated from a point source. For indoor operations, local exhaust (hood) ventilation is generally a more economical method of control when compared with dilution ventilation since the required air volume flow rate is substantially less. Local exhaust systems should be desgined to provide a capture velocity of 100 to 150 linear-feet-per min. in the direction of the exhaust inlet at the point of contaminant evolution. The exhaust inlet should be designed to insure contaminant capture. For this purpose, total enclosure is optimal but not always attainable. Thus, efforts should be made to enclose as much of the contaminant source as is practical. The location of the exhaust inlet should be selected to take advantage of the natural movement of the contaminant. The material, however, should not be allowed to pass through an individual's breathing zone enroute to the exhaust inlet. References compiled at the end of this chapter provide more in-depth information on laser technology and biological effects.

To estimate potential health hazards from exposure to laser radiation requires information on the following:

- the shape or profile of the laser beam intensity distribution
- how the profile changes as the beam traverses the atmosphere
- the defining aperture for the optical radiation protection standards

The beam profile at a fixed distance from a single-mode laser is approximately a Gaussian distribution. Mathematically, this distribution can be written in terms of the irradiance $E(r)$ as a function of radial distance r from the center axis of the beam:

$$E(r) = E_0 e^{-r^2/2\sigma^2} \qquad (5.6)$$

where E_0 = peak irradiance; σ = constant, related to the width of the distribution. Normally, the radiant exposure beam profile at the exit of a solid-state pulsed ruby laser system does not even remotely follow a Gaussian distribution. At great distances from the laser, however, the beam is "truncated" and broken up into various hot spots. This change in the shape of the beam occurs due to diffraction at the laser's projection optics as well as interactions between the beam and the atmosphere. Measurements of maximal beam irradiance at several points downrange permit the calculation of an effective beam diameter which can be related to σ. As such, Equation (5.6) can also be used for a pulsed laser system with beam radiant exposure $H(r)$ as a function of radial distance assuming an effective value for σ.

The laser beam diameter is not directly apparent for a Gaussian distribution as opposed to a rectangular beam profile (refer to Figure 5.11). The beam diameter can be defined such that the peak irradiance can be calculated. Consider the total power contained within a beam with radial symmetry. This total power, ϕ, is given by

$$\phi = \int_0^\infty E(r)\, 2\pi r\, dr \qquad (5.7)$$

where $2\pi r\, dr$ is the differential area of an infinite diameter circular aperture through which the beam passes. Combining the above expressions and integrating, gives

$$\phi = \pi D_\Delta^2 E_0/4 \qquad (5.8)$$

where $D_\Delta = 2\sqrt{2}\sigma$, D_Δ is twice the radial distance to where the irradiance on

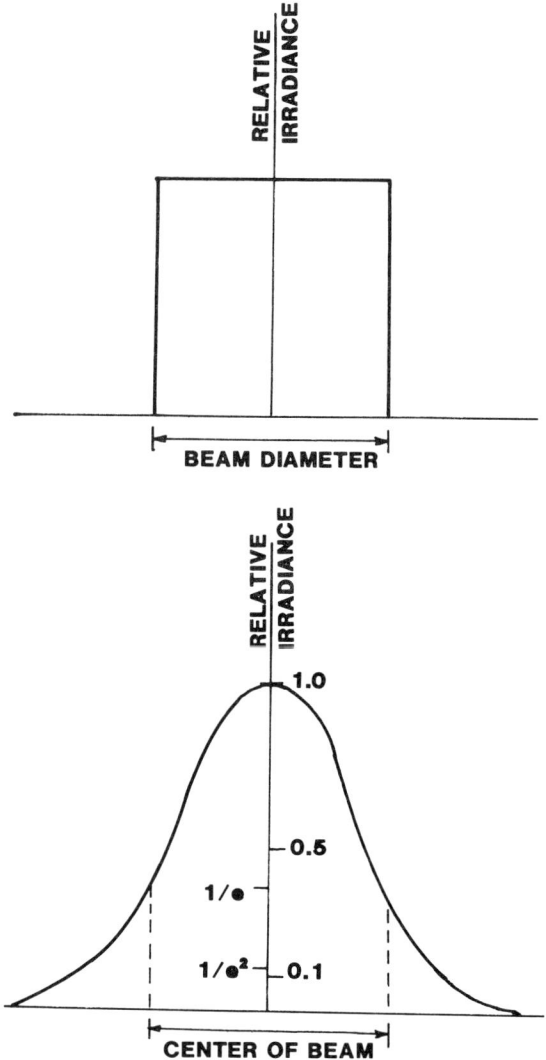

FIGURE 5.11. Shows Gaussian and rectangular beam profiles.

the Gaussian distribution is reduced to E_0/e (or beam diameter to $1/e$-peak-irradiance-points). From the beam diameter defined at $1/e$-peak-irradiance-points and the total power contained within the Gaussian profile it is possible to compute the peak irradiance with the same calculation as for a rectangular beam. One simple method for experimentally measuring the beam diameter consists of allowing 63% or $1-1/e$ of the total beam power to pass through an adjustable circular aperture located on the beam axis. The diameter of this aperture is D_Δ. Figure 5.12 provides a plot of this integral over various limits of integration.

The beam profile at any other point along its path will also be approximately Gaussian. The Gaussian beam in the far field will widen and the peak irradiance will be reduced, the farther the beam is from the laser. The total power within the beam will be reduced only slightly due to atmospheric absorption. The beam diameter at some distance, r, from the laser is

$$D_\Delta = r\tan\phi + a \qquad (5.9)$$

where ϕ = beam divergence and a = diameter to $1/e$, peak-irradiance-

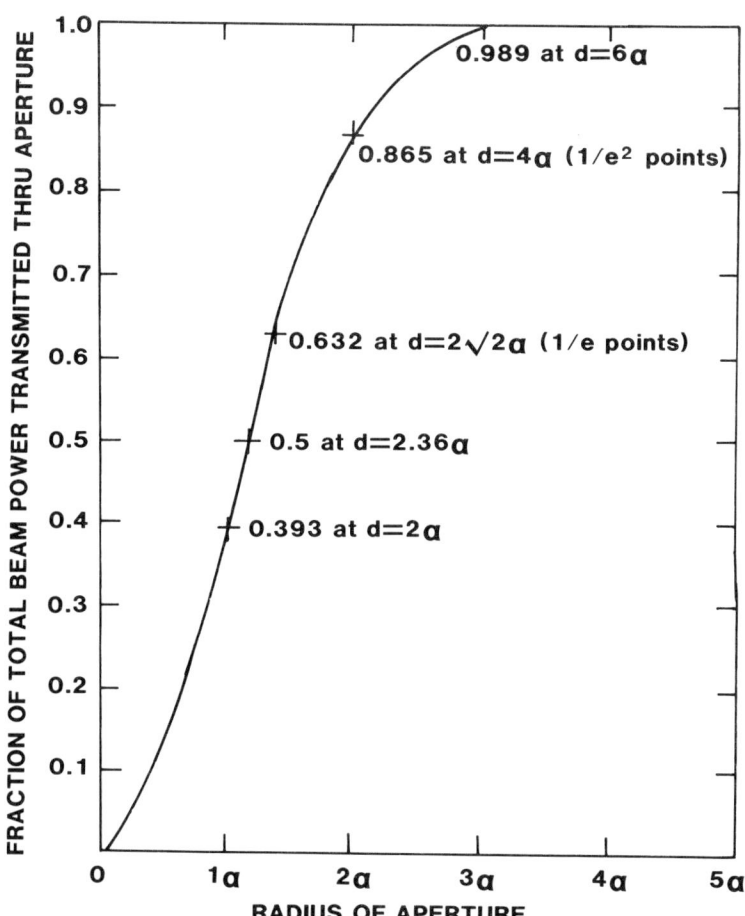

FIGURE 5.12. Beam diameter determined by measuring fraction of total power in Gaussian beam passing through a calibrated aperture.

points at the laser output. Since most laser systems are highly collimated, then

$$D_\Delta = r\phi + a \qquad (5.10)$$

From this relation we note that the beam divergence must also be specified to $1/e$-peak-irradiance-points. To demonstrate this, consider specifying the beam diameter to $1/e^2$-peak-irradiance-points (D'), then from Equation (5.6):

$$D_\Delta = D'/\sqrt{2} \qquad (5.11)$$

and

$$a = a'/\sqrt{2} \qquad (5.12)$$

therefore:

$$\phi = \phi'/\sqrt{2} \qquad (5.13)$$

where a' and ϕ' are the exit beam diameter and divergence, respectively, defined to $1/e^2$-peak-irradiance.

The laser range equation is obtained by combining Equations (5.8) and (5.10), or:

$$E(r) = (1.27\phi\, e^{-\mu r})/(a + r\phi)^2 \qquad (5.14)$$

where $e^{-\mu r}$ = atmospheric transmission. (μ is called the atmospheric attenuation coefficient and is normally very small.) Hence, although beam divergence could be defined in several ways, it is convenient for a hazard evaluation to select the beam divergence defined at $1/e$-peak-irradiance-points so that it is possible to predict the peak irradiance within a Gaussian profile at any distance from a laser of known output power. We can also apply this equation to experimentally measure the beam divergence. We can measure the peak irradiance with a detector whose sensitive diameter is much smaller than D_Δ for the beam in the far field of the laser ($r >> a$) and then compute ϕ from Equation (5.14) since ϕ, μ, r and a can also be measured.

It is important to note that usually the beam diameter, D_Δ, is much larger than the sampling diameter for the laser protection standards since the potential hazard often extends to great distances from the laser. The protection standards for the skin and cornea and lens of the eye are based upon power or energy transmitted through a 1-mm aperture (for wavelengths between 10^5 and 10^6 nm this aperture becomes 10 mm), whereas the aperture for the "retinal hazard region" of the spectrum (400–1400 nm) is based upon a 7-mm aperture (dark-adapted pupillary diameter). Actual range measurements are performed on laser systems with small exit beam diameters using an appropriate diameter aperture placed directly in front of the detector and centered on the beam axis.

The maximum power (energy available to pass through the appropriate defining aperture) is the most useful parameter to determine from the standpoint of evaluating optical radiation hazards. Figure 5.12 relates the fraction of total power transmitted through different diameter apertures when the total beam power is known. The Figure 5.12 curve states that the power through an arbitrary axial aperture of diameter d is

$$\phi d = \phi(1 - \beta) \tag{5.15}$$

where $\beta = E(d/2)/E_0$ and $d = 2\sqrt{2\ell n(1/\beta)}$. By integrating the Gaussian profile over the area of an arbitrary circular aperture and combining this power with Equation (5.8), we obtain

$$\phi d = \phi\left[1 - e^{(d/D_\Delta)^2}\right] \tag{5.16}$$

This expression can be applied to laser systems which have relatively short retinal hazardous ranges (D_Δ is of the same order of magnitude as 7 mm). The average irradiance over an arbitrary axial circular aperture of diameter d is given by

$$E(r,d) = 2.6\phi\left[1 - e^{(d/D_\Delta)^2}\right]e^{-\mu 4} \tag{5.17}$$

This range equation is primarily applied to low power Ga-As laser diodes and He-Ne lasers.

REFERENCES

1 American National Standards Institute. "Safe Use of Lasers," Standard Z-136.1, New York:ANSI (1973).

2 Charschan, S. S. *Lasers in Industry*. New York:Van Nostrand Reinhold Co. (1972).

3 Cheremisinoff, N. P. *Instrumentation for Complex Fluid Flows*. Lancaster, PA: Technomic Publishing Co. (1986).

4 Clark, A. M., W. J. Geeraets and W. T. Ham, Jr. *Appl. Opt.*, 8:1051–1054 (May, 1969).

5 Geeraets, W. J. and E. R. Berry. "Ocular Spectral Characteristics as Related to Hazards from Lasers and Other Light Sources," *Amer. J. Opthal.*, 66:15–20 (July, 1968).

6 Leibowitz, H. M. and G. R. Peacock. *Arch. Opthal.*, 81:713–721 (May, 1969).

7 Michaelson, S. M. "Occupational Health and Radiation Hazards," *Occupational Health and Safety*, pp. 28–37 (May/June, 1979).

8 Rockwell, R. J. and L. Goldman. "Research on Human Skin, Laser Damage

Threshold," Contract No. F41609–72–C–0007, prepared for the U.S. Air Force Aerospace Medical Div., Brooks Air Force Base, TX (June, 1974).

9 Skeen, C. H. et al. *Ocular Effects of Repetitive Laser Pulses*. Technology, Inc., San Antonio, TX, Air Force Contract F41609–71–C–0018 (June 30, 1972) (AD 7467695).

10 Sliney, D. H. and B. C. Freasier. "Evaluation of Optical Radiation Hazards," *Applied Optics, Vol. 12*. (Jan. 1973).

11 Sliney, D. H., D. C. Vorpahl and D. C. Winburn. *Arch. Environ. Health*, 30:174–179 (April, 1975).

12 Solon, L. R. *Science*, 134:1506–1508 (1967).

13 Stein, M. and S. Elgin. *Measurements of Retinal Image for Laser Radiation on Rhesus Monkeys*. USAF Contr. F41609–68–C0038, Eye Research Foundation of Bethesda (Feb., 1970).

14 Title 29, Code of Federal Regulations 1910, Occupational Safety and Health Administration (OSHA) Standards (Oct. 18, 1972).

15 U.S. Dept. of Health, Education and Welfare, Bureau of Radiological Health. "Laser Products, Performance Standards," *Federal Register*, 40(148):32252–32265 (July 31, 1975).

16 Wolbarsht, M. L. and D. H. Sliney. "The Formulation of Protection Standards for Lasers," Chapter 10 in *Laser Applications in Medicine and Biology, Vol. 12*. M. L. Wolbarsht, ed. New York:Plenum Press (1974).

PART II
Ionizing Radiation

6 / ENVIRONMENTAL RADIATION

PROPERTIES OF NATURAL RADIATION

Most terrestrial background radiation consists of gamma rays or X-rays, which largely result from the radioactive decay of unstable potassium, thorium, uranium and various other radioactive elements contained in the soil. These elements are widely distributed, with minerals containing them found around the world (refer to Table 6.1). X-rays are typically generated as a result of electrons being ejected from the atom's nucleus when it spontaneously disintegrates. Henri Becquerel discovered natural radioactivity in 1986 by radiographing a key. He found that the key, made of heavy metal, stopped most of the X-rays from exposing a film on a photographic plate. One year before, Wilhelm Roentgen generated X-rays artificially by focusing electrons onto a metallic target. Shortly after these discoveries it was determined that the atoms of such heavy metals as uranium and thorium are constantly, but slowly breaking down and emitting alpha rays, beta rays, as well as X-rays.

Alpha rays are streams of alpha particles which are positively charged. They are comprised of two neutrons and two protons (actually a helium nucleus). It is the least penetrating of the three common forms of radiation (alpha, beta, gamma). It is so weak that it can be stopped by a sheet of paper and is dangerous only when inhaled or ingested. Beta rays are streams of beta particles. These are identical with electrons when negative, and when positive, are called positrons.

External radiation exposure is associated with a steady rain of charged particles, or cosmic rays, moving at nearly the speed of light and falling upon the earth at all times and from all directions. These particles are just the nuclei of ordinary atoms stripped of their electrons—for the most part the nuclei of hydrogen (protons).

A large part of cosmic radiation consists of photons (or electromagnetic radiation) covering a fairly large spectrum of energies. Another part consists of

Table 6.1. Radioactive elements in the lithosphere.

Isotope	Abundance	Half-life (years)	Radiation
Radium-226	2×10^{-12} g/g*	1622	alpha, gamma
Uranium-238	4×10^{-6} g/g	4.5×10^9	alpha
Thorium-232	12×10^{-6} g/g	1.4×10^{10}	alpha, gamma
Potassium-40	3×10^{-2} g/g	1.3×10^9	beta, gamma
Vanadium-50	0.2 ppm**	5×10^{14}	gamma
Rubidium-87	75 ppm	4.7×10^{10}	beta
Indium-115	0.1 ppm	6×10^{14}	beta
Lanthanum-138	0.01 ppm	1.1×10^{11}	beta, gamma
Samarium-147	1 ppm	1.2×10^{11}	alpha
Lutetium-176	0.01 ppm	2.1×10^{10}	beta, gamma

*Gram per gram of soil.   **Parts per million.

subatomic particles, called <u>mesons</u>, whose mass is intermediate between the electron and the proton. Another small but important component is a flux or wash of high-energy neutrons. Neutrons are uncharged elementary particles with mass slightly greater than that of a proton. The isolated neutron is unstable and decays with a half-life of about 13 minutes into an electron and a proton. Neutrons sustain the fission chain reaction in a nuclear reactor. Half-life is the time in which half the atoms in a radioactive substance disintegrate to another nuclear form. The neutrons are eventually slowed down to the point where their energy is about the same as that of the molecules of our atmosphere, which is the same as saying that they end up at the same temperature. At this slow speed they are readily captured by nitrogen atoms to form radio-active carbon (C^{14}) or super-heavy radioactive hydrogen called tritium (H^3). Refer to Table 6.2 for a list of radioisotopes produced by cosmic rays. The dis-

Table 6.2. Some radioisotopes produced by cosmic rays.

Isotope	Half-life	Concentration (dis/min/cu.m)*
Tritium-3	12.3 yrs.	10^1
Beryllium-7	53 days	1
Beryllium-10	2.7×10^6 yrs.	10^{-7}
Carbon-14	5760 yrs.	4
Sodium-22	2.6 yrs.	10^{-4}
Silicon-32	700 yrs.	2×10^{-6}
Phosphorus-32	14.3 days	2×10^{-2}
Phosphorus-33	25 days	1.5×10^{-2}
Sulfur-35	87 days	1.5×10^{-2}
Chlorine-36	3×10^5 yrs.	3×10^{-8}

*Disintegrations per minute per cubic meter of air in the lower troposphere.

integration of these isotopes contributes little to the external exposure because their radiation is not penetrating. However, in gaseous form these materials can be ingested or inhaled and are dangerous if taken in large quantities. This brings us to the unavoidable problem of the intake of radioactivity from food and drink.

INTERNAL RADIATION EXPOSURE

Whenever the issue of internal exposure due to natural radioactivity in food is considered, Brazil often comes to mind. Not only because of the very high gamma-ray activity of the soil, which exists in certain areas of Brazil as well as India, but also because of the Brazil nut with its extraordinarily high radioactivity—some 14,000 times that of common fruits. Cereals are also relatively high—perhaps as much as 500–600 times that of fruits, which have the lowest concentrations of natural radioactivity. Table 6.3 lists some foods known to have high background radiation levels.

The table reports only the alpha activity of foods and this tells nothing about the radioactive potassium contribution. All muscle-building foods contain potassium. For most populations the radioactive potassium in our muscles provides 10–20 times the internal exposure of any of the other incorporated radioactive materials, such as radium or carbon-14.

Another source of internal exposure is well water. In the Midwest there is a high natural radioactivity in the drinking water. As early as 1955 the radioac-

Table 6.3. Relative alpha activity of foods.*

Food Stuff	Relative Activity
Brazil nuts	1400
Cereals	60
Teas	40
Liver and kidney	15
Flours	14
Peanuts and peanut butter	12
Chocolates	8
Biscuits	2
Milks (evaporated)	1–2
Fish	1–2
Cheeses and eggs	0.9
Vegetables	0.7
Meats	0.5
Fruits	0.1

*From *Proceedings of the Second United Nations International Conference on the Peaceful Uses of Atomic Energy*, September 1–13, 1958, Geneva, Switzerland, United Nations Publication, Volume 23, Experience in Radiological Protection, page 153, W. V. Mayneord.

tivity was found to result from radium, which had been leached out of the soil along with the other more common salts, such as calcium, magnesium, etc. The deep wells of Joliet, Illinois, for example, have 300 times more radium than Chicago's lake water. The water from Maine's wells has 3000 times the radium concentration of the Potomac River. This radioactivity is mild when compared with that of some springs in Boulder, Colorado, or Joachimsthal, Czechoslovakia, where the concentration is 10,000 times higher.

RADIOACTIVE EFFECTS AND OCCURRENCE IN NATURE

Radiation damage to the molecules within a cell is lethal. The delicate structure of the genes and chromosomes is particularly vulnerable to the impact of radiation, and broken chromosomes are the main cause of actual cell death. A cell that is not killed outright may be so damaged as to be unable to undergo replication (mitosis). In humans, there are many tissues whose cells must undergo division throughout life. Hair and fingernails grow as a result of cell division at their roots. The outer layers of skin are steadily lost through abrasion and are replaced through constant cell division in the deeper layers. The same is true of the linings of the mouth, throat, stomach, and intestines. Also, bloodcells are continually breaking up and must be replaced.

If radiation kills the mechanism of division in only some of these cells, it is possible that those that remain intact can divide and eventually replace or do the work of those that can no longer divide. In that case the symptoms of radiation sickness are relatively mild and eventually disappear.

Where radiation is insufficient to render a cell incapable of division, it may still induce mutations, and it is in this way that radiation may induce skin cancer, leukemia, and other diseases. It is this type of somatic or bodily cell damage that may arise from natural radiation background although the probability, like the background's intensity, is extremely low. Mutations can be generated in the sex cells also. When that occurs, succeeding generations are affected and not merely the exposed individual. The relatively mild effect of mutations is more serious than the drastic one of nondivision. A fertilized ovum that can't divide eventually dies and does no harm; one that can divide, but is altered, may give rise to a defective individual.

Experiments in the early 1930s by geneticist Hermann J. Muller and others showed that the number of mutations was directly proportional to the quantity of radiation absorbed. It is generally believed that this proportionality is valid at very low radiation doses; however, there is no threshold for the genetic effect of radiation. This is quite different from the somatic effect. A small radiation dose may affect growing tissue and prevent a small proportion of the cells of those tissues from dividing. The remaining, unaffected cells take up the slack, however, and if the proportion of affected cells is small enough, symptoms never become visible.

In the genetic effect, however, there is no undamaged cell that can take over the work of the affected sex cell once fertilization has occurred. Consider for example only one sex cell out of a million is damaged. It will take part, on the average, in one out of every million fertilizations. When it is used, it will not matter that there are 999,999 perfectly good sex cells that might have been used—it was the damaged cell that was used. That is why there is no threshold in the genetic effect of radiation and why there is no "safe" amount of radiation insofar as genetic effects are concerned.

Background radiation accounts for much less than 1% of the spontaneous mutations that take place naturally. The others arise out of chemical effects, random heat (molecular vibration) effects, etc. If the background radiation is increased by a factor of two or three, only a small portion of the mutation rate that radiation causes will be doubled or tripled. Thus, we can have a large factor of increase in the background radiation and yet only increase the total number of mutations by 1 or 2%.

The type of radiation we are exposed to is a critical parameter. Alpha rays, which barely penetrate the skin, can have no genetic effect when kept outside the body. Of course, if these were somehow ingested and deposited within the gonadic tissues, they could cause severe damage. Conversely, these same alphas could result in considerable local damage to skin tissue by depositing all their energy in a very thin layer. Thus, certain organs and areas of our body are more susceptible than others to radiations of various kinds. Any discussion of potential damage by radiation must deal with its composition as well as with its intensity. In radiobiology, particles can be compared on the basis of their average Linear Energy Transfer (LET), which is the average amount of energy lost per unit of distance traveled. To understand better the potential biological effects of radioactivity, we need first to gain a better understanding of its origin.

First, we note that the only reliable method of measuring such very long time intervals is based on the discovery by Pierre and Marie Cure in 1898 that some atoms are radioactive, i.e., they change spontaneously into other atoms at regular and constant rates. The original radioactive atoms are referred to as parent atoms, and the atoms they disintegrate into are daughter atoms. When we measure the population ratio of parent atoms to daughter atoms and perhaps even great-grandparents to great-granddaughters for very long-lived elements like uranium, one can obtain a good estimate of the age of the earth, which is 4.5 billion years.

To determine the age of man, we look for a natural radioactive material that decays at a much faster rate. When we find a piece of wood in some ancient structure, we can measure the amount of carbon in it, determine how much of it is C^{14} (half of whose atoms decay every 6000 years), and then calculate back to the time when the radioactivity from the C^{14} was the same as we now find in living wood, where it is continually being replenished from the atmosphere.

Natural radioactivity has also been very useful as a tracer in studying a great variety of geologic, meteorologic, oceanographic, and soil science problems. A measurement of the radon content of water samples is a measurement of its radium content. As the radon content of the atmosphere is negligible, compared to that in surface seawater, radon continually escapes from the sea. By analyzing the vertical distribution of radon deficiency in the surface ocean it is possible to determine the rates of both vertical mixing and gas exchange. Water trapped in deep-sea sediments contains 10^4–10^5 times more radon than the overlying seawater. Hence, radon diffuses from the sediment into the sea. The vertical distribution of excess radon in near-bottom water provides an index of the rate of vertical mixing.

The radium-226 content of surface water in both the Atlantic and Pacific Oceans is uniformly close to about 4×10^{-14} gram per liter. The deep Pacific has a concentration of radium-226 that is four times higher, and the deep Atlantic has a concentration twice as high as that of the surface. These distribution profiles can be explained by the same particle-settling rate for radium-226 from surface to depth for the two oceans and by a threefold longer residence time of water in the deep Pacific than in the deep Atlantic.

A common, naturally occurring radioactive isotope is lead-210, which undergoes radioactive decay to become polonium-210. Lead-210 in cigarette smoke is the primary source of the increased concentrations of both these radioelements in body tissues. These higher concentrations, in turn, may increase by 8–30% the internal radiation dose to the bones of cigarette smokers over that received by mankind as a whole.

Safety engineers need to know the kinds and quantities of natural radiation to which man is exposed. Before a reactor-powered electricity generating facility is constructed on a site, public relations people usually call in radiation safety officers to make measurements of the natural background. This then provides a threshold or base against which to monitor future plant operations. Thus, if the effluent cooling water exhibits more beta-ray activity than was present in the intake from the local stream, measures should be taken to improve the purification technique. Of course, the local public health people might insist that the water coming out of the plant be cleaner than when it went in.

Such stringent and excessive restrictions on the part of the public authorities are based on the political facts of life, namely, a psychological fear of radioactivity. Another factor that is of great concern to reactor safety people is the possibility of discharging radioactivity into the air. The general method used in sampling the atmosphere is to draw air through a filter at a known rate for a specified time interval. The filtered radioactivity is then measured and the activity per unit volume determined. A problem with this technique is that there are days when the air is heavy with the daughters of radium: radon, radium A, B, and C. The concentration of these natural and short-lived radioisotopes can be as much as 500 times the maximum permissible concentration for exposure of the general population to long-lived plutonium-239, which may have escaped from the reactor. It behooves the radiation protection officer to be aware of natural contaminants, their properties, and potential quantities so that the proper detective work will be carried out on the monitoring data.

RADIATION DETECTION

From the first discovery, studies of radiation effects on matter continued to be carried out. Exclusively solid matter, crystalline particularly, was studied in the early laboratories. It was soon learned that the passage of radiation through certain chemicals, like zinc sulfide (ZnS), would give rise to flashes of light that could be seen with the naked eye. Lord Ernest Rutherford and his students spent hours in a darkroom looking into a small eyepiece coated with ZnS, which he called a scinthariscope. They observed and counted the individual flashes of light or scintillations resulting from the impact of alpha rays. Today we are able to amplify these light flashes electronically and record both their number and intensity. In this way we can measure the rate of exposure fairly easily, whereas the emulsion can only sum or integrate the radiation over a period of time.

Other solids, crystals, and glasses are capable of storing radiation energy. When interrogated by heating or ultraviolet irradiation, these solids give off visible light that is proportional to the stored radiation energy. An application that is most suitable for such solid, luminescent dosimeters is the evaluation

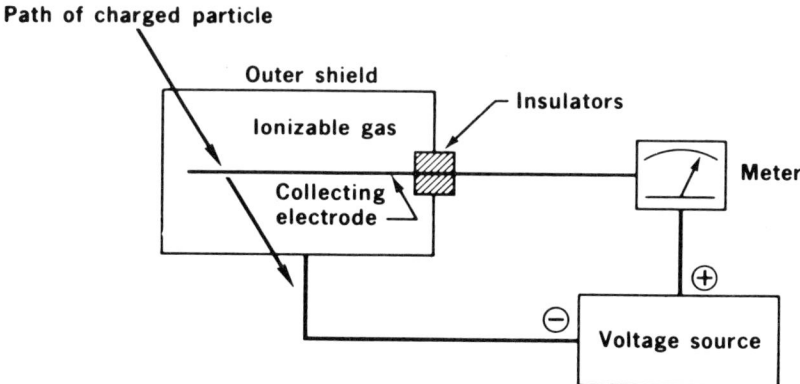

FIGURE 6.1. Gas ionization chamber.

of the natural radiation environment as related to plant growth; this is essentially a study in ecology. The advantage of such solids is their relative resistance to weathering, small temperature changes, humidity, etc.

As laboratory techniques became more sophisticated and physicists learned how to create vacuums, they began to study gases. Some of these experiments were the basis for the discovery of X-rays. The ionization (or the splitting into charged fragments) produced in a gas is one of the most common and sensitive indicators of the passage of radiation. Figure 6.1 shows a typical set-up for the detection of radiation. Radiation in the form of charged particles strikes the ionizable gas in the little chamber (formed by the outer shield) and causes ions to be formed. If the voltage is high enough the ions will be attracted to the shield or to the collecting electrode, depending on their charge, before they can recombine. A little current (or pulse) will flow for each radiation particle that interacts in this way, and these pulses will be measured (or counted) on the meter. If the voltage is increased, more ionization occurs when the initial ions collide on the way to the electrodes. This is called gas amplification, and the pulses are much larger although still proportional to the ionization energy delivered by the radiation particle. This detector is called a proportional counter and it measures small ionizations.

If the voltage is increased further, the pulses become very large and alike in size. This type detector is the Geiger-Muller counter, which, although very sensitive, tells little about the kind of radiation that has been encountered.

7 / ATOMIC ENERGY *AND* ITS BIOLOGICAL EFFECTS

DEFINITIONS AND CONCEPTS

The smallest complete unit of matter is the atom; the next larger is the molecule. All molecules are made up of atoms. All matter is composed of molecules, which can be very simple, or extremely complicated. There are many different kinds of atoms and a molecule may be made up of a cluster of one kind only or a cluster of different kinds. Salt, for example, is a very simple molecule which has two kinds of atoms sticking together to form the molecule. When the molecule contains two or more kinds of atoms we have a mixture or "compound" type of matter. If it contains one kind only, we have a pure element or elemental substance.

The constitution of matter is always a vast number of molecules, each molecule having a fixed arrangement of atoms, whether matter is solid, liquid or gas. Water is always H_2O (two parts hydrogen, one part oxygen), whether it comes as ice, or from the hot or cold faucet, or as steam. The forms in which we find matter (solid, liquid, gas) are called states of matter.

Molecules are not packed close together. Rather, they have space between them. The more dense a substance is, the less space there is between its molecules. Gases have the most space between their molecules, liquids have less space, and solids the least of all. But the molecules of all matter are constantly in motion. In gases, the speed of movement is tremendous. In liquids, the speed is reduced. In solids, the movement may be described as a slight vibration. In other words, everything is movement. Table 7.1 lists the elements.

When the molecules making up a substance are moving fast, the temperature of that substance is high; when the molecules are slowed down, the temperature of that substance decreases. Sometimes the speed of molecules in liquids and gases becomes so great (through heating, for example) that the molecules at the surface can overcome the cohesive forces holding them together. They then fly off into space. This is characterized by boiling or evaporation.

Table 7.1. List of elements.

Number of Protons	Name of Element	Symbol of Element	Discoverer of Element	Where Discovered	When Discovered	Weight of Element
1	Hydrogen	H	Cavendish	England	1766	1
2	Helium	He	Ramsay	Scotland	1895	4
3	Lithium	Li	Arfvedson	Sweden	1817	7
4	Beryllium	Be	Wohler	Germany	1828	9
5	Boron	B	Davey	England	1807	11
6	Carbon	C			Ancient	12
7	Nitrogen	N	Rutherford	Scotland	1772	14
8	Oxygen	O	Priestley	England	1774	16
9	Fluorine	F	Moissan	France	1886	19
10	Neon	Ne	Ramsay and Travers	Scotland	1898	20
11	Sodium	Na	Davey	England	1807	22
12	Magnesium	Mg	Davey	England	1808	24
13	Aluminum	Al	Wohler	Germany	1827	27
14	Silicon	Si	Berzelius	Sweden	1823	28
15	Phosphorus	P	Brand	Germany	1669	31
16	Sulfur	S			Ancient	32
17	Chlorine	Cl	Scheele	Sweden	1774	35
18	Argon	A	Ramsay and Rayleigh	Scotland	1894	39
19	Potassium	K	Davey	England	1807	39
20	Calcium	Ca	Davey	England	1808	40
21	Scandium	Sc	Nilson	Sweden	1879	45
22	Titanium	Ti	Gregor	England	1791	48
23	Vanadium	V	Sefstrom	Sweden	1831	51
24	Chromium	Cr	Vaquelin	France	1797	52
25	Manganese	Mn	Gahn	Sweden	1774	55

(continued)

Table 7.1. (continued).

Number of Protons	Name of Element	Symbol of Element	Discoverer of Element	Where Discovered	When Discovered	Weight of Element
26	Iron	Fe			Ancient	56
27	Cobalt	Co	Brandt	Sweden	1735	59
28	Nickel	Ni	Cronstedt	Sweden	1751	59
29	Copper	Cu			Ancient	64
30	Zinc	Zn	Marggraf	Germany	1746	65
31	Gallium	Ga	Boisbaudran	France	1875	70
32	Germanium	Ge	Winkler	Germany	1886	73
33	Arsenic	As			Ancient	75
34	Selenium	Se	Berzelius	Sweden	1817	79
35	Bromine	Br	Balard; Lowig	France	1825	80
36	Krypton	Kr	Ramsay	Scotland	1898	84
37	Rubidium	Rb	Bunsen	Germany	1861	85
38	Strontium	Sr	Davey	England	1808	88
39	Yttrium	Y	Gadolin	Finland	1794	89
40	Zirconium	Zr	Klaproth	Germany	1789	91
41	Columbium	Cb	Hatchett	England	1801	93
42	Molybdenum	Mo	Scheele	Sweden	1778	96
43	Technetium	Tc	Noddack and Tacke	Germany	1925	99
44	Ruthenium	Ru	Klaus	Russia	1844	102
45	Thodium	Rh	Wollaston	England	1803	103
46	Palladium	Pd	Wollaston	England	1803	107
47	Silver	Ag			Ancient	108
48	Cadmium	Cd	Stromeyer	Germany	1817	112
49	Indium	In	Reich and Richter	Germany	1863	115
50	Tin	Sn			Ancient	119
51	Antimony	Sb			Ancient	122
52	Tellurium	Te	Muller	Rumania	1782	128

(continued)

113

Table 7.1. (continued).

Number of Protons	Name of Element	Symbol of Element	Discoverer of Element	Where Discovered	When Discovered	Weight of Element
53	Iodine	I	Courtois	France	1813	127
54	Xenon	Xe	Ramsay	Scotland	1898	131
55	Cesium	Cs	Kirchoff and Bunsen	Germany	1861	133
56	Barium	Ba	Davey	England	1808	137
57	Lanthanum	La	Mosander	Sweden	1839	139
58	Cerium	Ce	Hisinger and Berzelius	Sweden	1805	140
59	Praseodymium	Pr	Welsbach	Austria	1885	141
60	Neodymium	Nd	Welsbach	Austria	1885	144
61	Promethium	Pm	Hopkins et al	United States	1926	147
62	Samarium	Sm	Biosbaudrar	France	1879	150
63	Europium	Eu	Demarcay	France	1901	152
64	Gadolinium	Gd	Boisbaudran	France	1886	157
65	Terbium	Tb	Mosander	Sweden	1843	159
66	Dysprosium	Dy	Boisbaudran	France	1886	162
67	Holmium	Ho	Cleve	Sweden	1879	165
68	Erbium	Er	Mosander	Sweden	1843	167
69	Thulium	Tm	Cleve	Sweden	1879	169
70	Ytterbium	Yb	de Marignac	Switzerland	1878	173
71	Lutecium	Lu	Urbain	France	1907	175
72	Hafnium	Hf	Coster and von Hevesey	Netherlands	1923	179
73	Tantalum	Ta	Ekeberg	Sweden	1802	181
74	Tungsten	W	Scheele	Sweden	1781	184
75	Rhenium	Re	Noddak and Tacke	Germany	1925	186
76	Osmium	Os	Tennant	England	1804	190
77	Iridium	Ir	Tennant	England	1804	193

(continued)

114

Table 7.1. (continued).

Number of Protons	Name of Element	Symbol of Element	Discoverer of Element	Where Discovered	When Discovered	Weight of Element
78	Platinum	Pt	Scalinger	Italy	1557	195
79	Gold	Au			Ancient	197
80	Mercury	Hg			Ancient	201
81	Thallium	Tl	Crookes	England	1861	204
82	Lead	Pb			Ancient	207
83	Bismuth	Bi	Geoffroy	France	1753	209
84	Polonium	Po	Curies	France	1898	210
85	Astatine	At	Allison	United States	1931	211
86	Radon	Rn	Dorn	Germany	1900	222
87	Francium	Fr	Allison and Perey	United States	1939	223
88	Radium	Ra	Curies	France	1898	226
89	Actinium	Ac	Debierne	France	1899	227
90	Thorium	Th	Berzelius	Sweden	1828	232
91	Protoctinium	Pa	Hahn and Meitner	Canada	1917	231
92	Uranium	U	Klaproth	Germany	1789	238
93	Neptunium	Np	McMillan and Abelson	United States	1940	239
94	Plutonium	Pu	Seaborg et al	United States	1940	239
95	Americium	Am	Seaborg et al	United States	1945	241
96	Curium	Cm	Seaborg et al	United States	1944	242
97	Berkelium	Bk	Seaborg et al	United States	1949	245
98	Californium	Cf	Seaborg et al	United States	1950	246
99	Einsteinium	E	U. S. Team	United States	1952	253
100	Fermium	Fm	U. S. Team	United States	1952	256
101	Mendelevium	Mv	Seaborg et al	United States	1955	256
102	Nobelium	No	International Team	Sweden	1957	253–254?

Note: Elements 1 through 92 occur normally in nature. Elements 93 and beyond are those discovered by man as a result of transmutation.

115

The more than 400,000 substances known in the world today are made up of various combinations of more than 100 elements. When atoms of an element join together to form molecules, they share electrons. Thus, all chemical actions and reactions are caused by interchange and rearrangements of the electrons which revolve around the nuclei of the atoms involved.

All chemical reactions are the transferring of atoms from one molecular arrangement to another. Some bonds are tighter than others. The tighter the bond, the more stable the compound. When atoms change from a looser to a tighter arrangement, they give off energy in the form of heat, electricity, etc.

When atoms change from a tighter to a looser arrangement, energy from outside is required in the form of heat or electricity. (Example: carbon and oxygen make up carbon dioxide gas. To separate the gas back into its two components, energy in the form of heat at high temperature must be supplied.)

When a current of electricity is sent through a tube which has been made almost a vacuum, and when certain phosphorescent materials are put in the path of the current, small bursts of light can be seen. These bursts are caused by small particles hitting the phosphorescent material. These particles are electrons and their moving constitutes an electric current.

When atoms join together to form molecules, they share electrons. Thus, all chemical arrangements and actions are caused by electrons on the surface of atoms.

As noted earlier, the atom is made up of the nucleus, composed of protons and neutrons (with the exception of the simple hydrogen atom which has only one proton), around which electrons swirl. Electrons are arranged in neat and predictable patterns around the nucleus, revolving around the nucleus the way the planets revolve around the sun. The electrons are arranged in successive "shells" or orbits. Hydrogen has one shell, with one electron revolving in it. Helium has one shell with two electrons in it. Neon has two shells with two electrons in the inner shell and eight in the outer. Argon has three shells with two electrons in the inner, eight in the middle, and eight in the outer. The disturbance of electrons in the outer shells of atoms produces light rays which have long waves; disturbance of electrons in the inner shells produces rays of shorter wave length, such as X-rays.

The atomic number of an element is determined by the number of protons in its nucleus. Hydrogen has one proton, so it is number 1. The number of protons in helium is two, which is the same as its atomic number. However, its weight is four. The weight or mass of an atom is determined by adding together the number of protons and the number of neutrons in the nucleus. Thus, helium has two neutrons and two protons. Neutrons and protons are the same weight. Each weighs one atomic unit. It takes six hundred sextillion (6×10^{23}) atomic units to make a gram.

Electrons weigh almost nothing and so are disregarded when weight is considered. There are always the same number of electrons revolving around the nucleus as there are protons in the nucleus.

The greater the weight of an element, the greater are the number of protons and neutrons in the nucleus. The atomic weights of the elements are given in Table 7.1.

Physicists have shown that the mass, or weight, of each atom (except ordinary hydrogen) is less than the total mass of the protons, neutrons and electrons which compose it. That is, the whole is less than the sum of its parts. To understand it, two scientific principles have to be put together. The first principle is the conservation of matter: matter can be neither created nor destroyed, but only altered in form. The second principle concerns conservation of energy: energy can be neither created nor destroyed, but only altered in form. Since some mass in the nucleus of the atom disappears and since mass cannot be destroyed, it must be converted to energy.

Albert Einstein suggested as early as 1905 that mass and energy are two forms of the same thing. He reasoned that just as energy can be converted to various forms, such as heat and electricity, and just as matter can be converted to various forms, such as solids, liquids, and gases, so can mass and energy be converted to each other.

Out of this reasoning came the famous Einstein formula. He showed that to translate mass into energy, mass has to be multiplied by the square of the speed of light. Thus, the formula is $E = MC^2$, energy equals mass times C (for speed of light) squared. (Energy is expressed in ergs, mass is expressed in grams and the speed of light in centimeters per second.) Since the speed of light is about 186,000 miles per second, you can see that a very little mass converted would yield a great deal of energy.

Protons and neutrons which are free are in an unstable state. When they go together into a nucleus they are in a more stable state, but they lose a slight amount of mass. This mass is converted to energy, some of which binds the protons and neutrons together in the nucleus, and some of which is released.

The difference between the mass of the free protons and neutrons before joining and the mass of the nucleus which they form is called the packing loss. The energy which holds the protons and neutrons together is called the binding energy.

Elements which have the same chemical properties cannot be separated by chemical means. The chemical properties are determined by the number of electrons. But there are atoms of elements which have the same chemical properties but different atomic weights. The difference in weights, the difference in their nuclear mass, is accounted for by the fact that the nuclei have the same number of protons but a slight difference in the number of neutrons. These atoms are called isotopes, the word coming from the Greek *isos* meaning "same" and *topos* meaning "place." Figure 7.1 gives examples of isotopes.

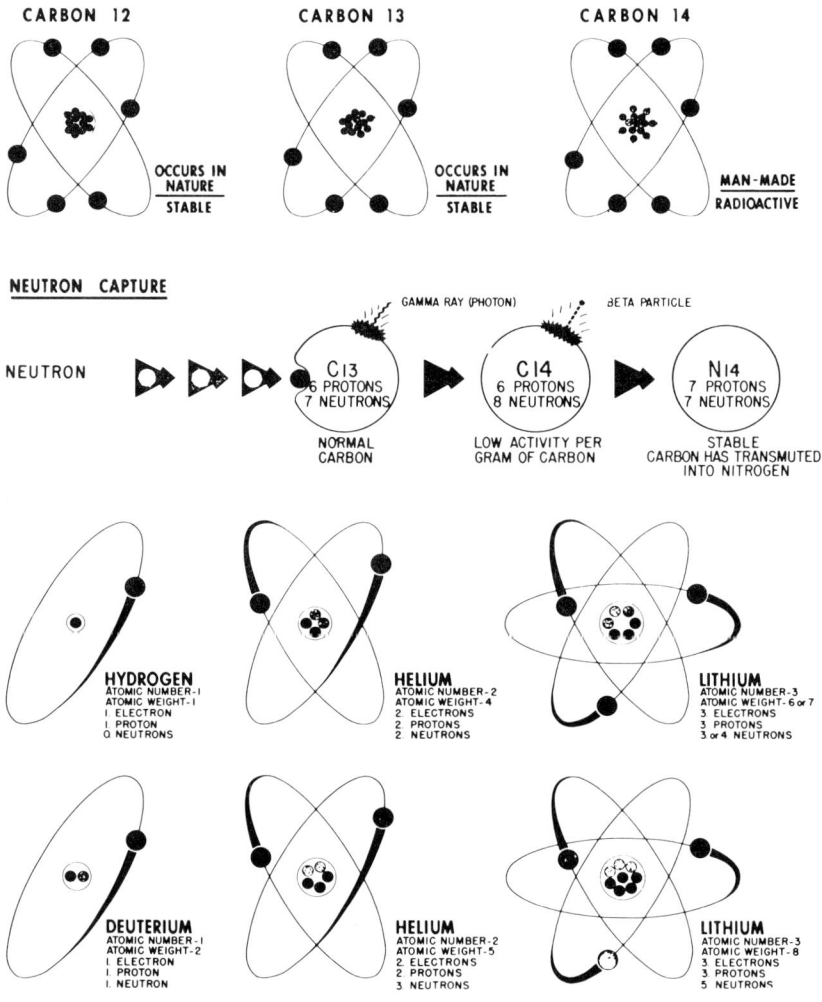

FIGURE 7.1. Isotopes.

All atoms have isotopes, there being more than 1,000 from the elements we know. A few of the natural isotopes are radioacive—they emit rays of radiation, but most of the natural isotopes are stable. Man, having developed new tools in the form of atomic reactors, or atomic furnaces, in which the nuclei of elements can be altered by bombardment of neutrons, can now make his own radioisotopes, changing stable elements into unstable ones. The latter give off energy in the form of rays.

PRINCIPLES OF ATOMIC ENERGY

Consider what happens when a small piece of carbon-13 is placed in an atomic reactor. Left within the reactor for a period of a week or a year, depending on how much of the radioactive form of carbon (C^{14}) you want, it is then removed. Once removed, it is now radioactive—it is giving off energy in the form of rays—because some of the carbon-13 atoms have picked up an additional neutron and become agitated. The nuclei of the C^{13} atoms were struck by neutrons generated in the reactor. These neutrons stuck within the nucleus and changed its complexion.

The extra neutron puts some additional energy into the nucleus which it has to get rid of. This energy is released by the neutron breaking down into a proton and a beta particle (an electron) being emitted. (Theoretically, a neutron may be assumed to be composed of a proton and an electron.) Once this electron, or beta particle, is emitted, the radioactive carbon atom changes to a new element—nitrogen which has 7 protons (one more than carbon) and 7 neutrons. Figure 7.2 illustrates nuclear radiation.

Uranium, naturally radioactive, occurs in nature in three forms—it has isotopes, which are variants of an element with different weights.

The uranium isotopes are:

U-234 = 92 protons and 142 neutrons (weight 234);
U-235 = 92 protons and 143 neutrons (weight 235);
U-238 = 92 protons and 146 neutrons (weight 238).

In its natural state, almost 99.3 percent of uranium occurs as U-238. Seven-

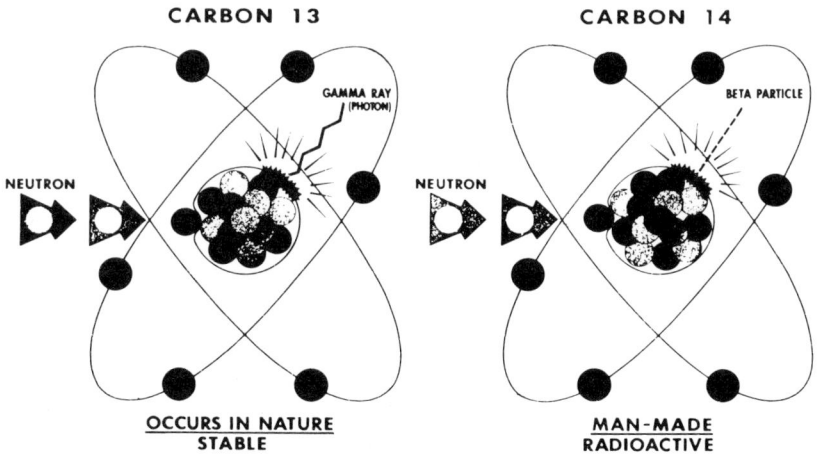

FIGURE 7.2. Nuclear radiation.

tenths of one percent (0.7%) occurs as U-235. The U-234 occurs in such trace amounts that it is barely discernible.

In general, there are two classes of fuels for atomic energy. One type is made up of heavy atoms like those of U-235. Energy is released when the atoms are split into smaller atoms roughly equal in size. The other type of fuel consists of the lightweight atoms, such as hydrogen, which through a certain mechanism may be made to fuse or combine with one another to form new atoms.

Fission means a breakup into parts, or a splitting; fusion means bringing things together.

So far as we know today, the uranium isotope 235 is the only one in nature that will undergo fission when hit by a neutron. The fissioning effect of this particular isotope had not been scientifically observed prior to 1939. Under proper conditions, uranium-235 fissions or splits with release of great amounts of energy.

Uranium-235 and a man-made element, plutonium, are the basic atomic fuels. The source material for both uranium-235 and plutonium is normal uranium. The uranium-238 mentioned above cannot be used directly as an atomic fuel but under certain conditions in an atomic reactor or furnace can be converted to plutonium-239, the man-made element which has fissionable qualities just like uranium-235. In brief, uranium-238 is not a spontaneous atomic fuel but in an atomic furnace can be converted to an element which is an atomic fuel.

U-235 can be separated directly from U-238 by physical methods that depend on the slight difference in the weights of their atoms. The large ratio of 140 atoms of U-238 to each atom of U-235 in the natural uranium ore also complicates the process of producing high purity U-235.

A stray neutron which is free in the atmosphere and traveling at a high rate of speed will eventually burst into the nucleus of the U-235 atom. The uranium nucleus then splits or fissions, usually into two fragments. These fragments are new elements, such as barium and krypton, which are about one-half the weight of uranium. Two or more neutrons fly off into space at the same time. One neutron releases two or more others, and if there are more uranium nuclei in the path of the released neutrons, the same reaction will take place again. Each of two neutrons will split another uranium atom, forming more barium and krypton and two or more neutrons, so there are now at least four neutrons. This process continues, resulting in a chain reaction.

The number of neutrons varies from fission to fission but for the sake of simplification we will consider three neutrons for some fissions. With each fission, the increasing number of neutrons represents an increasing amount of energy—power. Figure 7.3 illustrates the process.

The release of neutrons in ever-increasing numbers in the fission of billions of trillions of atoms in the arrangement of U-235 in a very small fraction of

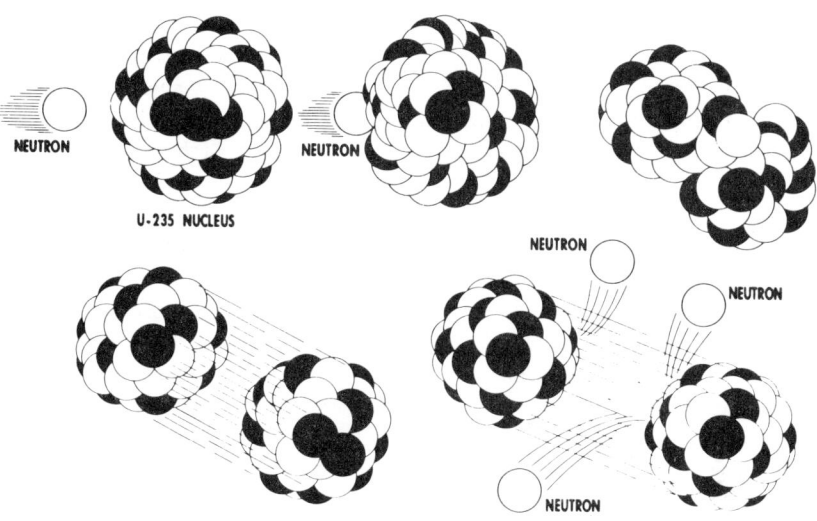

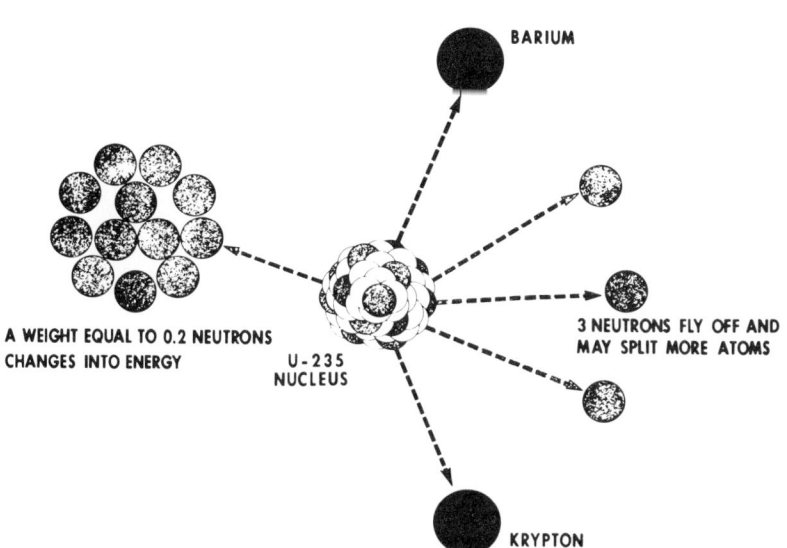

FIGURE 7.3. Fissioning of uranium-235.

a second is known as a chain reaction. The fission processes and the chain reaction throughout the fuel of an atomic bomb are practically instantaneous. So, in an unbelievably short period of time, the fuel has reacted and one has a large variety of fission product atoms that are intensely radioactive. We also have a large number of free neutrons because the expanding chain ran out of fuel.

These neutrons are very penetrating. They radiate and travel from the center of the explosion with high velocities. Neutron radiation can affect plants and animals and many inert materials. Simultaneously, during the progress of the chain reaction, radiant energy in the form of gamma rays, light, and heat are released from each of the billions of trillions of atoms that undergo fission. The cumulative heat radiated is sufficient to start fires and cause burns at some distance. The gamma rays travel with the speed of light, are highly penetrating, and can severely damage living organisms.

If a piece of atomic fuel is relatively small, the probability exists that the neutron will pass out of the fuel entirely and be lost from the chain reaction. If the amount or shape of the atomic fuel is such that the neutrons escape too easily and do not cause sufficient new fissions, then the chain reaction will die. Such an assembly is said to be subcritical. Therefore, the size and shape of the fuel mass must be such that one will get on the average more than one additional atom to split for each atom that has been split. This fuel assembly is said to be supercritical. An assembly in which fission by continuous chain reaction is proceeding at a constant rate under control is said to be critical This is what happens in an atomic reactor.

There is a method by which a system of atomic fuel can be kept subcritical and therefore stable, ready to be converted to a supercritical condition.

Two pieces of fuel are kept subcritical and separated until one desires to bring them together. In the subcritical state, there is a great deal of surface through which the neutrons can escape. If too many escape, a chain reaction, even if started, would soon die out due to the loss of too many neutrons without causing fission. By bringing together these two subcritical pieces, part of the loss is checked because neutrons now flow between the two pieces. Although there is some loss of neutrons at the surface of the mass, the combination makes it possible to increase the amount of fission in each generation, giving a supercritical assembly and an explosion.

A reactor, in chemical engineering terms, is simply a piece of equipment or pressure tank designed to contain, or provide for, chemical reactions between chemical substances, i.e., reactions involving the electrons of the particular substances. A nuclear reactor is one in which reactions within the nuclei of atoms occur in a self-sustaining, or chain-reacting manner. As pointed out previously, a nuclear reactor operation is a controlled fission process.

The amount of concentrated atomic fuel necessary for an atomic bomb can be measured in pounds, but it is possible to get a continuous chain reaction to take place in the U-235 of normal uranium if we properly assemble tons of the diluted material. (A ton of uranium is small in size when compared to a ton of coal, since uranium is the heaviest naturally occurring element.)

A pile is a nuclear reactor, or atomic furnace. Figure 7.4 illustrates the action of a uranium-graphite pile. The initiating neutron works its way through the graphite (carbon) atoms, striking a uranium nucleus at A, and causing its fission. Two more neutrons are emitted. The neutron following path (1) collides with graphite nuclei. These collisions reduce its speed until it causes the disintegration of another uranium nucleus at B. The disintegrating nucleus gives two more neutrons. The neutron following path (3) passes through the uranium mass C without causing disintegration, and proceeds to D, where it causes further reaction. The neutron proceeding along path (2) is captured by the uranium nucleus at E, since it is moving at a relatively high speed. The neutron following path (4) escapes through the surface of the pile. Figure 7.5 shows the reactor components.

The basic event in a nuclear reactor is the breaking up of nuclei of U-235 atoms, i.e., uranium atoms having a weight of 235 on the atomic scale. Of course, an atom of uranium has a cloud of electrons surrounding its nucleus just as other heavy atoms do, but these electrons are unimportant in the opera-

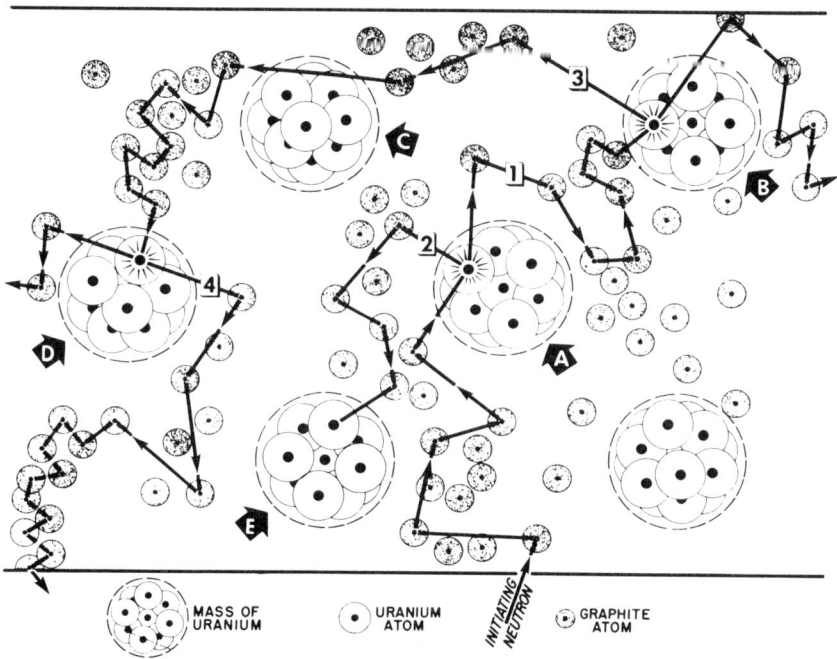

FIGURE 7.4. Operation of a uranium-graphite pile.

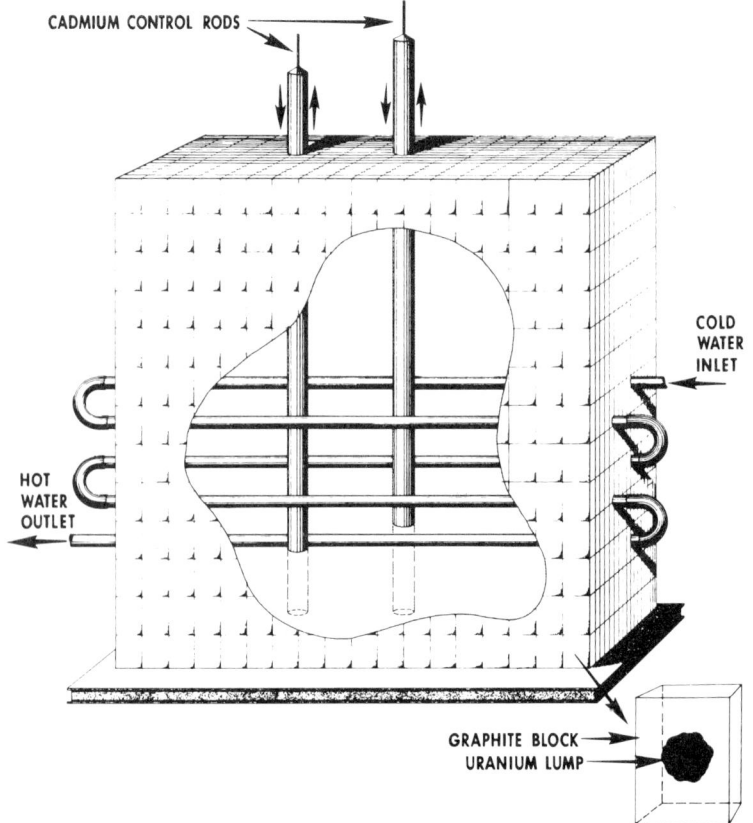

CADMIUM CONTROL RODS

COLD
WATER
INLET

HOT
WATER
OUTLET

GRAPHITE BLOCK
URANIUM LUMP

FIGURE 7.5. Components of a uranium-graphite pile reactor.

tion of the reactor and may be ignored. The U-235 nucleus can be broken up if struck by a slowly moving neutron that happens to be roving about, originating, say, from a cosmic ray.

When a neutron of the right speed strikes a U-235 nucleus, a series of events take place:

1. The nucleus splits into two unequal parts that fly in opposite directions at extremely high speed.
2. Two or three neutrons get loose from the main pieces and start out independently.
3. A lot of energy is released as heat.
4. Gamma rays and other forms of radioactivity are produced.

The neutrons which originate in the fission and start out independently may go to any one of three places. First, one must cause the fissioning of at least one more U-235 nucleus if the chain reaction process in a block of uranium is to continue. They will not do so, if they escape from the piece of uranium, which they are very likely to do, since a neutron is small in comparison with the distance between atom centers (nuclei) in the solid metal.

Secondly, and of equal importance, some may be captured by the nuclei of U-238 atoms.

The capture by the heavier isotope, U-238, is highly probable, because there are so many of them in uranium in comparison to the U-235 variety. There are no fewer than 140 U-238 nuclei for every U-235 nucleus. These two kinds of nuclei behave quite differently when struck by a neutron. Thirdly, some neutrons escape the uranium and if not reflected back, will be captured by other materials and lost to the process. The main idea is to keep this loss as low as possible and make the most of the neutrons that strike the uranium-238 or 235, since both these events are necessary to the process.

If a neutron that has been liberated during fission of a U-235 nucleus immediately strikes another U-235 nucleus, it is slowed somewhat by the encounter but only jolts the heavy nucleus a bit. However, if it has collided with so many nuclei that its speed has been reduced from an original value of about 30,000,000 miles per hour to about 5000 or lower, it will cause fissioning or splitting of any U-235 nucleus it may then happen to hit.

If moving at high speed, a neutron bounces off a U-238 nucleus much as it would from one of U-235. However, if it has been slowed by successive collisions to an intermediate speed of 10,000 to 100,000 miles per hour, it is likely to stick to a nucleus of U-238. Thus, that particular neutron is lost for producing fissions of U-235 and sustaining the chain reaction, but still performs a useful purpose.

Addition of one neutron to a U-238 atom changes it to U-239 which is unstable (radioactive). Neptunium-239 is formed from the U-239. The neptunium-239 is likewise unstable and it decays to plutonium-239, which is more stable. Briefly then, plutonium-239 is formed from uranium-238 sometime after the uranium-238 picks up a neutron in an atomic reactor. The plutonium, although a different chemical element, is fissionable the same as U-235. It can be left in the reactor to help sustain the chain reaction or it can later be separated chemically from the unchanged uranium and the fission products. The plutonium-239 can then be concentrated and replaced in the reactor as an atomic fuel in the same manner as uranium-235.

There are several components in a reactor. The reactor must first have an atomic fuel, uranium-235, which is introduced as a small part of the tons of normal uranium in the reactor. The uranium metal is usually in the form of metal rods or slugs which must be sealed in thin-walled aluminum cans to protect them from oxidation, since they heat up when the reactor is in operation.

Second, a moderator must be used to slow down the neutrons given off by uranium-235 fission so that the chain reaction can continue efficiently. Slow neutrons are much more effective in causing uranium-235 nuclei to fission than are fast ones in the presence of the large amount of uranium-238. The moderator is usually a material having lightweight atoms which the neutrons can hit and thereby lose part of their motion. Two of the best moderating materials are heavy water (D_2O) and carbon in the form of graphite. Ordinary water is also used extensively.

Another component is the control- and safety-rod mechanism for control of the reaction. If the chain reaction were allowed to continue to expand, the reactor could conceivably be put out of commission through a burn-up and melting down of materials.

Atoms of some materials have a strong ability to capture free neutrons and hold them. If such a material in the form of rods is placed in a reactor, it can completely stop the chain reaction by capturing the neutrons, thus making them unavailable to the fuel in sufficient amounts to propagate the chain. Boron and cadmium are effective materials for capture of neutrons. Rods containing either of these can be moved in and out of the reactor and positioned to allow the chain reaction to operate at the desired level of reaction.

If a reactor is operating at a constant level, it can be speeded up to a higher level by merely backing the control rods out of the reactor. After the reactor has reached the higher level, the rods are replaced in their original position to hold that level of operation. The operating level can be decreased to a new lower level or the reactor can be completely shut down by merely pushing the control rods farther into the reactor.

To start a reactor that has been completely shut down, it is only necessary to adjust the position of the control rods; there are always a few free neutrons available to restart the chain reaction. A reactor also can be started by being primed with a radioactive element which throws out a great many neutrons.

In operating a reactor to produce plutonium-239, it is desirable to have it operate at as high a power level as possible in order to speed up the rate of production. However, the higher rates of operation mean higher rates of fission of uranium-235; this in turn means greater amounts of heat generated.

Cooling the reactor is thus necessary. Some reactors are cooled by air; some by water; and others by liquid-metal organic compounds and other fluids. The coolant keeps the reactor at sufficiently low temperature to avoid heat damage to the materials of construction, and, in the case of power reactors, to convey the heat for the production of steam and the operation of turbo-generators, which, in turn, produce electricity.

When a reactor is in operation many neutrons and gamma rays are emitted and pass out through its surfaces. Due to the damaging effects of these radiations to living tissues, it is essential that the reactor be properly shielded to protect the operators.

The shield may be made of iron and concrete several feet in thickness. This shield completely encloses the reactor when it is in operation. Portholes or removable panels in the shield afford access to the reactor.

After a charge of uranium has been in a production reactor long enough to convert the proper amount of uranium-238 to plutonium-239 the canned rods are then discharged from the reactor. Chemical treatments remove the unchanged uranium and fission products and concentrate the atomic fuel, plutonium-239. The fission products, or ashes from the atomic fuel, are extremely radioactive and complicate these chemical treatments by making it necessary to carry out the processing by remote control behind heavy shields. In a power reactor, the plutonium may be left in for long periods of time and burned as fuel. Eventually, of course, the old fuel must be removed and replaced with a fresh supply. Reactors are used for production of electricity, of materials, and for atomic weapons, research, and production of radioactive elements that can be used beneficially in medicine, industry, and agriculture. In a reactor, atomic energy develops in the form of heat; a reactor is merely a new type of heat source. To obtain electric power, a heat exchange medium or coolant must be used as a gas or fluid to transfer the heat. The heat is transferred to ordinary water in a heat exchanger or boiler to form high pressure steam to turn a turbine and generator. But to the usual engineering problems of high duty heat transfer are added problems peculiar to reactor operation, such as a level of dangerous radioactivity and temperatures so high that the usual building materials, like mild steel, will not stand up under operation.

USES OF RADIOACTIVE MATERIALS

No other field has benefited as profoundly from radioisotopes as the life sciences. As tracer atoms, these tools are uniquely suitable to solution of problems in this field. Radioactive tracers correspond in size and kind to the things they trace. Radioisotopes are "natural" tracers and, as far as biological processes are concerned, are no different from other atoms except in their ability to send out signals to identify their presence.

Useful radioactive forms are available for most elements which enter into biological processes. Sodium, for example, is important in body fluids and tissues and is easily traced by means of its radioisotope, sodium-24. Transfer of sodium to various body sites is traced and measured by means of the sodium-24 gamma rays. Only a few seconds are required for the blood stream to carry the sodium from one arm, through the heart and lungs, and into the other arm.

Some processes in the body are so complex that only through radioisotope traces has any success in unraveling them been achieved.

Of the 92 natural elements, 15 or 20 are known to be essential to life processes. Most of these have been used in tracer studies designed to show their normal metabolism in the body. For example, radiotracer studies showed that nearly 90% of injected calcium became concentrated in the bones of young animals; in older animals the bone uptake was about 40%. Dietary factors have been found which reduce the amount of calcium available to the body. This action may possibly be of value in treating certain diseases. Even more effective is the compound ethylenediaminetetraacetic acid. With a great affinity for calcium, it tends to remove it from the body.

A large part of the work with metabolites is motivated by the hope of cancer control. If differences can be found between the needs of cancer cells and those of normal tissue, it may be possible to retard or prevent the growth of one without seriously affecting the other.

A large and important number of man's diseases are those which affect the blood. Among these are anemia and the cancer-like conditions, leukemia and polycythemia vera. The advent of radioactive iron has provided a powerful means of studying the production and makeup of blood.

Radioisotopes are invaluable in studies of cholesterol and its role in arteriosclerosis. The loss of elasticity and the thickening of arteries and capillaries which characterize this condition are caused in part, at least, by the deposition of cholesterol in fatty accumulations on their inner surfaces.

Cholesterol metabolism is now better understood through tracer studies. C-14-labeled acetate fed to rats soon results in the appearance of C-14 cholesterol in many tissues of the body, mostly in the blood and liver. Rates of rise and fall in C-14 activity at the various sites give clues to its production and travels. The rise in activity in the blood is found to be later than in the liver; this and other considerations show that the liver is the site of blood cholesterol production.

Many diagnostic applications are based on measurement of the volume or amount of a substance in the body by means of the isotope dilution technique. It frequently becomes necessary to determine a patient's blood volume. In a typical application, human serum albumin labeled with I-131 is injected into the blood stream. After sufficient time for mixing (10 to 15 minutes), a sample of blood is withdrawn and measured for radioactivity. The degree to which the label has become diluted then gives a measure of the blood volume.

In many abnormal conditions, such as edema, the volume of total body water and of water in intracellular and extracellular spaces must be determined. Dilution techniques provide the answer. Total water is measured from the dilution of water with deuterium or tritium. The mixing of a tracer amount throughout the body and in the cells is essentially complete in about an hour, and a dilution sample can be taken at that time. No convenient method exists for measuring volume within the cells, but this can be found as the difference between total body water and extracellular water. The latter is measured by means of a tracer. Certain substances with large molecules, such as insulin and sucrose, have been used for the purpose but faster mixing in the extracellular fluid is obtained with radioactive tracers such as sodium-24 and bromine-82.

Knowledge of the heart's output is invaluable in diagnosis of many heart disorders. As with other pumps, the efficiency of the heart can be judged by the flow it induces. To measure this output, a small amount of human serum albumin labeled with I-131 is injected into an arm vein. The resulting peak of activity in the blood is measured after it is pumped through the heart and then passes a certain point in an artery. The ability to trace blood circulation and accumulation at particular sites with radioactive materials has been valuable also in cases of hemorrhage, burns, and fistulae between organs as in stab wounds.

Radioisotopes of phosphorus, sulfur, calcium, copper, molybdenum, and zinc are being used to investigate important plant problems. Radiotracers are being used to determine the path, rate of entry, and movement of material in plants, and the mechanism of such movement. Tracing element through living plants was impossible before radioisotopes became available. One of the most extensive uses of isotopes in agriculture concerns studies on uptake of fertilizers. With radioactive tracers it is possible to study both commercial fertilizers and green manures.

With radioisotopes, it is possible to tag a certain material such as phosphorus in the fertilizer, and tell how much moves into the plant. The effectiveness of different methods of placement of the fertilizer such as below the seed, broadcast, or foliage application can be compared. It has been proved that plants absorb nutrients through the foliage, the fruit, the twigs, the trunk, and the flowers, as well as through the roots. Tracer studies show that certain plant nutrients applied to the leaves in soluble form result in as much as 95% uptake by the plant. Plant nutrients applied to the soil often show no more than 10% uptake.

Maintenance of soil fertility is the main objective of soil management. Soils of the United States corn belt are fertile, but their continued ability to supply adequate potash (potassium) to corn has been questioned. The capacity to supply potash can be measured with radioisotopes. Unfortunately, one cannot use a radioisotope of potassium in field studies for it has none with long enough half-life. However, radioactive rubidium, which behaves chemically similarly to potassium, can be used and follows along with potassium as a tracer.

Calcium-45 has also been used to study the absorption of calcium by alfalfa and other legumes as influenced by soil reaction, amount of exchangeable calcium, degree of calcium saturation, presence of manganese, iron and aluminum, and method of application.

A major portion of the radioisotope applications in the animal husbandry field pertains to metabolism studies in poultry, cattle, and sheep. Radioactive sulfur has been used to study selenium poisoning. Selenium is taken up by plants and, when consumed by animals, has many toxic effects. Sulfur metabolism was shown to be affected by the ingestion of selenium. The latter is believed to interact with sulfur in important amino acids, thus making them unavailable to the body. Now a better insight has been gained into the action of selenium. Other studies include the use of radiophosphorus to investigate the metabolic relation of phosphorus with iron, cobalt, copper, and molybdenum in larger animals. Mineral metabolism in animals such as rabbits, chickens, swine, and cattle has been studied with the following radioisotopes: cesium-134, phosphorus-32, sulfur-35, calcium-45, iron-55, cobalt-60, copper-64, zinc-65, strontium-89, molybdenum-99, iodine-131, and tantalum-182.

There are many other examples, such as insect studies, migration and hibernation, and genetics.

The ability of radiation to penetrate matter has led to extensive industrial use of radioisotopes. Industrial radiography is one of the large uses of radioisotopes. Radiography with radium, of course, has been used to inspect metal castings and welds for possible flaws. Reactor-produced cobalt-60, cesium-137, and iridium-192 are now used in the same way as radium but are cheaper and more effective.

In another application, the transmission-type thickness gage is used on production lines to help produce more uniform paper, aluminum, copper, tin plate, plastics, rubber, glass, and numerous other items. This continuous, noncontacting method of gaging is especially useful where products are moving rapidly, where temperatures are high, and where products are soft and may be easily marred.

In the transmission-type of thickness gage, the amount of radiation picked up by the detector depends upon the amount of material between it and the radioactive source. In such measurements, it is assumed that the density of the material remains constant. Conversely, if the thickness or size of a material remains constant, the same gage may be used to measure density. Examples are the monitoring of the density of a process fluid flowing through a pipe or the moisture content of sand or other material.

Radiation reflection assures accurate measurement of thickness of coatings laid over a base metal. For example, it is possible to measure the thickness of gold as plated over copper or rubber over steel. Soil density and soil moisture gages for use in the field are other industrial applications. They have been used in the selection of suitable sites for aircraft runways, highway roadbeds, and hydroelectric dams. Each apparatus uses a radioactive source, a radiation detector, and an electronic recorder. The source and detector are housed in a probe which can be lowered through a one-inch-diameter steel tube driven into the ground. The soil density probe utilizes a cobalt-60 source. Gamma rays bombard the soil surrounding the tube and are scattered by it. Some return to the Geiger counter which is mounted in the top of the probe and shielded against direct radiation from the cobalt-60. The amount of reflected radiation can be translated directly into soil density in pounds per cubic foot.

The ability of atomic radiation to ionize most of the materials it strikes can be put to many beneficial industrial uses. For example, static electricity in industry is a menacing problem. It occurs quite generally whenever products possess insulating properties. Through their ability to ionize the air at selected points along the moving stock and thus "ground" the static electricity, radioisotopes help to meet this problem.

It has been demonstrated that beneficial chemical and physical reactions may occur when certain materials are subjected to the proper amount of radiation. Many new compounds, and, especially, some superior plastics, have already been made. Breaking down oil through radiation promises better gasoline.

Gamma radiation in large enough doses can destroy bacteria and enzymes in a material without raising its temperature significantly. The recent availability of large sources of radiation from reactors has increased interest in radiation sterilization. The absence of high temperatures is necessary in the processing of many items such as antibiotics and certain other drugs. Large-scale sterilization or pasteurization of perishable foodstuffs such as meats, bananas, potatoes, and beverages holds interesting possibilities. Cold radiation sterilization has great potential importance.

Another example is the use of radiosodium or radioiodine to detect leaks in water lines. A small quantity of the radioisotope is introduced into the line and the path of its radiation followed until it reaches the spot where the radiation reading drops off. These tests have made it possible to find and repair leaks in buildings with a minimum disruption of the structure. Quick and reliable results are obtained where other techniques fail.

Radioisotopes provide a means of preventing what is called <u>color soiling</u> in multicolor textile printing operations. This occurs whenever one color is carried forward by the fabric from one printing roller to the next, which ruins valuable fabric. The offending or so-called pirate color can be labeled with radioactivity, and its gradual invasion of a sensitive color dye box carefully and continuously monitored. It is not necessary to synthesize a costly radioactive dye for this use. A few millicuries of phosphorous-32 as soluble phosphate are added to the dye bath in question.

THE PROBLEM OF RADIOACTIVE WASTES

The greatest source of wastes is the nuclear fuel cycle: the mining, milling, and preparation of fuel for reactors and weapons produce wastes containing natural radioisotopes; and fuel irradiation and subsequent processing produce wastes rich in fission products. Additional wastes are produced by irradiation of nonfuel materials in and around reactors.

Natural radioactivity refers to the radioactivity of materials found in nature. These materials are present in uranium and thorium ores, nature's nuclear fuels. Wastes from mining, ore milling, and fuel fabrication therefore contain this natural radioactivity, which consists primarily of the natural radioisotopes of uranium, thorium, radon (a gas), and radium.

Most mines are in dry formations, but some must be kept dry by pumping. The water that is removed contains only traces of radioisotopes and poses no significant health hazard. In uranium ore-concentrating mills, liquid wastes are discharged to ponds or lagoons at a rate of 300 to 500 gallons per minute (for an average mill processing 1000 tons of ore per day). Radium is the principal radioisotope in mill wastes and is incorporated into the solid residue, or tailings, in an insoluble form.

In feed preparation, the next step in the fuel cycle, ore concentrates are taken from the mills and chemically purified. This produces uranium salts, or feed, for the gaseous-diffusion plants (for separation of uranium-235 from other isotopes of uranium). or it converts concentrates to metallic uranium or uranium oxide for fuel elements. For each ton of uranium processed, approximately 1000 gallons of liquid wastes are produced.

The purified uranium used in the fabrication of fuel elements has extremely low activity because the radium, thorium, and several radioactive decay products have been removed in earlier steps of the cycle. Liquid wastes from fuel fabrication plants are of small volume and very low radioactivity. Contaminated scrap is also produced at fabricating plants.

Fission products manufactured during the irradiation of nuclear fuels in reactors are by far the largest source of radioactive waste in terms of contained radioactivity. When each uranium atom fissions, it breaks into two major fragments appropriately called fission products. Fission products are radioactive; they undergo one or more steps of radioactive decay before reaching a stable, harmless condition. Valuable fuel material remains in the irradiated fuel along with the accumulated fission products; so recovery of this fuel is important. Chemical processing of irradiated nuclear fuels is, therefore, an inescapable part of the nuclear industry. The processing creates highly radioactive wastes consisting not only of fission products but also of some activated reactor materials, chemicals, and corrosion products.

A number of solvent-extraction processes are used to separate remaining fuel from waste products in used fuel elements. From 1 to 100 gallons of highly radioactive liquid result from each kilogram of uranium processed—1 to 10 gallons in processing natural or slightly enriched uranium and 10 to 100 gallons in the case of highly enriched uranium. Table 7.2 provides a summary of the types of radioactivity.

Activation products are generated during the irradiation of nonfuel materials located near the fuel in nuclear reactors. Structural materials are activated (made radioactive) by the absorption of neutrons, as are impurities in the coolant and often the coolant itself. Traces of iron, nickel, and other corrosion products, for example, are carried along with the coolant in some types of reactors and irradiated as they pass through the reactor.

Some activation products are gaseous. In water-cooled reactors gaseous products arise from the irradiation of the water as well as from irradiation of any air in the coolant. The gases are mostly short-lived radioisotopes, such as nitrogen-16 or argon-41, but tritium (hydrogen-3) has a half-life of more than 12 years. Air-cooled reactors produce argon-41. Control and disposal of gases is an important aspect of radioactive waste management even though it involves a minor portion of the total radioactivity produced by nuclear industry.

Miscellaneous sources of radioactive wastes are the more than 4000 establishments making or using nuclear products in the United States. The radioactivity of this great variety of wastes originates in one of the ways we have just discussed — from naturally occurring radioisotopes, fission products or activation products. These may be in many forms, including chemicals, solids collected in evaporators, resins, and contaminated equipment and materials. Laboratories in hospitals, universities, and private industry produce waste solutions, contaminated gloves and clothing, and broken glassware. Even when sealed radioisotope devices are used, there eventually is need for disposal when the radioactivity has decayed below useful levels.

Radioactive wastes vary widely in the concentration of radioactive materials. It is convenient to classify them according to their potential hazard, as indicated by their level of activity. Three levels are defined.

Low-Level Wastes have a radioactive content sufficiently low to permit discharge to the environment with reasonable dilution or after relatively simple processing. They have no more than about 1000 times the concentrations considered safe for direct release. In liquid form, low-level wastes usually contain less than a microcurie of radioactivity per gallon.

Intermediate-Level Wastes have too high a concentration to permit release after simple dilution, yet are produced in relatively large volumes. Their radioactivity is approximately 100 to 1000 times higher than that of low-level wastes. In liquid form they may contain up to a curie of radioactivity per gallon.

High-Level Wastes contain several hundred to several thousand curies per gallon in liquid form and result from chemical processing of irradiated nuclear fuels. High-level wastes pose the most severe potential health hazard and the most complex technical problems in management.

Any discharge of radioisotopes to our environment launches a series of interrelated physical and biological events because of the movement of released materials through the atmosphere and underground water and their retention in soil and living organisms. Some of these processes may result in the concentration of radioisotopes that were originally released in very dilute form — thus fish may concentrate phosphorus-32 in their tissues, and oysters may accumulate zinc-65.

Radiation can affect an exposed individual directly (somatic effects) by damage to body cells, or it can affect his descendants (genetic effects) by damage

Table 7.2. Summary of types of radioactivity.

Type of Waste radioactivity	Source of waste	Form of waste	Typical isotopes	Type of radiation	Disposal methods
Natural activity	Mining of uranium ores	Solids	Uranium-238	α, γ	Pile in open
		Liquids	Thorium-230	α, γ	Seep into ground
		Gases and dusts	Radon-222	α	Ventilate mine
	Fuel fabrication plants	Solids	Uranium-238 Uranium-235	α, γ	Decontaminate
		Liquids (acid)	Uranium-238 Uranium-235	α, γ	Neutralize, concentrate, and bury residue
		Dusts	Uranium-238 Uranium-235	α, γ	Ventilate, filter, and disperse to air
Fission-product activity	Fuel irradiation and processing	Solids (from purposeful solidification	Strontium-90 Cesium-137	β β, γ	Encase in container and store permanently ($\sim$ 600 years)
		Liquids (with strontium and cesium removed)	Technetium-99 Ruthenium-103 Cerium-144	β β, γ β, γ	Store in tanks for several years; then solidify in place
		Gases	Iodine-131	β, γ	React with chemicals to bind in solid, e.g., silver iodide
			Krypton-85	β, γ	Disperse to air

(continued)

Table 7.2. (continued).

Type of Waste radioactivity	Source of waste	Form of waste	Typical isotopes	Type of radiation	Disposal methods
Activation-product activity	Reactor materials unavoidably irradiated during operation	Solids	Aluminum-28 Manganese-56	β, γ β, γ	Package and ship for land burial
		Liquids (dissolved materials	Cobalt-58	β^+, γ	Evaporation or ion-exchange; bury residue
		Gases	Nitrogen-16	β, γ	Hold for decay (very short life); then disperse to air
	Purposeful irradiation to produce useful isotopes	Solids	Cobalt-60	β, γ	Ship for burial when no longer useful (long life)
			Phosphorus-32	β	Store for decay to safe levels (short life)

to reproductive cells. Somatic effects are reduced when a given radiation exposure is spread over a long time. Although there is some evidence that the same principle applies in the case of genetic effects, it is not conclusive.

The isotopes of greatest concern are generally those with the longest half-lives, such as strontium-90 and cesium-137. It is also necessary to consider how each radioisotope is taken into the body and how it behaves, once there. Iodine-131, for example, is passed efficiently through food chains—from the air to grass, thence to cows, via milk to man, where it is concentrated in the thyroid gland. Radioactive iodine therefore requires stringent control even though its half-life is a relatively short eight days.

The type and energy of emitted radiation and the chemical properties of radioisotopes are other factors that were considered in establishing maximum permissible concentrations for the various radioisotopes. These safe limits are published in the Code of Federal Regulations, which controls management of radioactive waste materials.

Radioactive waste management consists of the various steps necessary for the safe disposal of radioactive wastes and the control and direction of these measures. Each type of waste requires its own methods for control and disposal. There are, however, two basic principles that are broadly applied.

Radioactive materials can be stored safely in permanently controlled reservations, but the volume of stored wastes would be prohibitively great if they were not first concentrated. Even high-level wastes can be concentrated or solidified for long-term storage.

Wastes of appropriately low activity may be reduced to permissible levels for release by dilution in air or in waterways. Wherever materials are to be released to the environment, the amount of radioactivity that can be safely dispersed into that particular environment is determined quantitatively for each specific radioisotope.

In the application of these principles, two considerations are important.

Each factor affecting waste management must be evaluated for each facility. These factors include accurate data on the specific radioisotopes, their chemical form and concentration, their maximum allowable concentration for release, and the detailed characteristics of the environment. Suitable treatment and disposal methods then can be chosen and detailed operating standards established.

It is important to make certain that operating standards for waste management are met. This requires constant checking on the amount and level of activity handled and released. If conditions change at a site, operating standards must be reexamined and changed if necessary.

Conscientious application of these considerations has been responsible for the success, safety, and economy of waste-management activities.

Liquid wastes of low and intermediate radioactivity generally have low solids content but vary considerably in chemical and radioisotope content. Treatment and disposal methods differ among nuclear installations, depending both on the waste content and the environment.

Small laboratories, such as educational, industrial, hospital, and medical laboratories, generally produce liquid wastes of low activity. They typically contain only traces of a few isotopes. They are produced in modest quantities and can usually be released by dilution, treated by filtration or ion exchange, or retained for shipment to sites equipped for their disposal. Though the amount of waste from these laboratories is small, all agencies licensed for use of radioactive materials must meet disposal standards.

Uranium mines and ore mills produce relatively large quantities of low-level liquid wastes requiring minimal treatment. Where mine pumping is required, the radioactivity is so low that the wastes are allowed to seep into nearby desert soil. The liquid wastes from ore-refining mills are also dispersed into the environment. Solids are settled out in large retention ponds, and the liquid overflows to streams or percolates into the ground. In some locations, mills first remove dissolved radium by chemical treatment.

Fuel fabrication plants produce modest quantities of low-level acid wastes that are diluted, neutralized, stored to permit decay, and then discharged to waterways. Some industrial fuel manufacturers collect the wastes and ship them elsewhere for disposal. All shipments must conform to Interstate Commerce Commission regulations, and approved containers must be used.

National Laboratories of the AEC produce many kinds of wastes. The methods used to manage them differ in detail but are similar in many respects. In general, liquid wastes from the laboratories range in concentration from less than a hundredth of a microcurie to several hundred microcuries per gallon. Treatment usually includes several of the following steps: filtration, chemical precipitation, ion exchange, evaporation, solidification in concrete, or absorption in porous materials such as vermiculite.

Chemical precipitation is a process of adding chemicals that combine with dissolved radioactive materials to form solid particles. The solids are then separated from the liquid and disposed of by packaging and burial. Chemical precipitation is 60–75% effective in removing mixed fission products from low-level laboratory wastes. Ion-exchange resins remove 90–99.99% of the contained radioactivity. Evaporation of low-level waste retains some 99.99% of the activity in evaporator concentrates.

Nuclear power plants process low-level liquid waste through treatment and storage systems, such as evaporators, ion exchangers, gas filtration, decay holdup tanks, incineration and fixation of solids and liquids in concrete. In many cases, liquid wastes are processed by decay storage without any other treatment. The quantities of radioactivity discharged from nuclear power plants have been well below permissible release limits. Most reactor installations control their wastes in accord with the strict standards applying to unidentified mixtures rather than less stringent ones for individual radioisotopes.

No liquid wastes are released into the ground at reactor installations. The waste systems do use decay hold-up tanks, evaporators, ion-exchange systems, filtration, and, in some cases, steam stripping (to remove gases). In a typical water-cooled reactor power plant, there are two sources of liquid waste: the reactor coolant itself and drainage from supporting laboratories and facilities. When necessary, contaminated liquids are processed in evaporators or ion exchangers. The evaporator concentrates and used ion-exchange resins are then retained for later shipment from the site. Ion exchange is typically used for continuous cleansing of reactor coolants, but discharged reactor water must nonetheless be retained in tanks to allow radioactive decay, or be processed further.

Gaseous wastes may contain natural radioactivity, fission-product activity, or materials activated by neutron absorption. In the mining, milling and fabricating of uranium into fuel elements, airborne radioactivity consists of natural radioisotopes in dusts or gaseous radon. These typically occur in low concentrations. Normal ventilation of the work area gives adequate protection. Air discharged from mine ventilation systems usually contains radon-222 and its short-lived decay products; the quantity released to the atmosphere is much less than that naturally given off by rocks of the earth's crust.

The generation of gaseous radioactive wastes at nuclear power plants is different for each type of reactor, but the wastes have been effectively managed in all types of reactors. In general, gaseous wastes are held for an appropriate period of decay and then are released through a high stack after filtration.

Radioiodine is often present in gaseous reactor wastes. Its presence usually determines how the wastes are managed. Activated charcoal an remove up to 99.9% of the radioiodine, but the charcoal's effectiveness soon can be reduced by surface contamination.

Fuel processing plants control the discharge of radioiodine by a combination of methods. First, the spent (used) fuel is stored from 90 to 120 days to allow natural decay of most of the iodine. Second, most remaining iodine is vaporized during the fuel processing and is removed from other gases by activated charcoal or a chemical-reaction system that converts the iodine into a solid.

Most solid radioactive wastes are of moderately low level and are disposed of by burial. As with liquids and gases, they may contain natural, fission-product, or activation-product radioactivity.

Solid refuse from mining operations contains only a fraction of 1% of uranium oxide. It is usually piled near mine portals; no hazard is involved because of its low radioactivity. Solid refuse from ore-refining mills, viz., tailings, is of greater concern because it contains most of the radium from the ore. The tailings are usually held in controlled areas to prevent dispersal to the environment. This is more economical than chemical treatment to remove the radium from the large quantities involved. How best to dispose of tailings actually remains an unanswered question for the present.

Solid wastes from feed-material and gaseous-diffusion plants contain only small amounts of natural radioisotopes. They are disposed of by burial in supervised reservations. Some are first treated to remove uranium, especially if it is enriched uranium.

Moderately radioactive solid wastes from laboratories and reactor installations are similar. They consist mostly of residues (used ion-exchange resins or evaporator concentrates) and contaminated equipment and materials (worn-out clothing, filter elements, glassware, blotting paper and other debris). Some establishments burn them to reduce their volume before disposal. Most, however, simply bale and bury them.

Land burial has been used extensively in the past for solid waste produced. Burial sites are carefully selected, and arrangements must be made to control wastes over a period of many years after they are buried. Land requirements for waste burial have averaged about 1 square foot of ground for 1.5 cubic feet of waste. Solid wastes are now being pressed in standard baling machines before burial, however to reduce the waste volume 2 to 10 times.

High-level wastes from processed nuclear fuels create the most challenging problem in waste management. These wastes produce substantial amounts of heat for a number of years. Furthermore, their long-lived radioisotopes require hundreds of years to decay to safe levels; during all this time they must be stored away from man and his environment.

Tank storage has been used since high-level waste management first began years ago at plutonium-production facilities. Tank storage of liquid wastes has proved to be both safe and practical. To extend this for hundreds of years will require periodic replacement of tanks, however, and no valid basis yet exists for accurately predicting tank service life. Experience indicates a reasonable tank-life expectation of several decades, however. Extensive research and development is being carried out to convert highly radioactive liquid wastes to a solid form.

Some installations reduce storage requirements by concentrating wastes, and improved fuel processing has reduced waste volumes from several thousand to several hundred liters per ton of spent fuel.

In-tank solidification calls for solidification of "nonheating" wastes by evaporation of the liquid. "Self-heating" wastes, on the other hand, will be separated into two fractions. The first fraction, containing the long-lived heat-producing radioisotopes, will be absorbed on ion-exchange beds and stored in stainless-steel cylinders; the remainder will become "low-heating" waste after several years and can be solidified in place in the underground tanks. Heated air (800°C) both heats and circulates the waste, and most of the water is boiled off. Upon cooling the remaining wastes form a massive salt cake that no conceivable material failure or natural phenomenon could expose to living plants or animals. Other methods of solidifying high-level wastes also are under development.

The disadvantages of tank storage for hundreds of years have resulted in vigorous search for more practicable methods for the ultimate disposal of high-level wastes. Two general approaches are being explored. The first seeks to convert aged wastes to solid form. The second seeks to store these solids deep

Table 7.3. Naturally occurring radioisotopes encountered in mining, milling, and fuel preparation in uranium fuel cycle.

Radioisotope	Atomic number	Percent found in natural uranium	Half-life	Radiation emitted*
Uranium-238 Decay Chain				
Uranium-238	92	99.28	4.5×10^9 yr	α, γ
Thorium-234	90	1×10^{-9}	24.1 days	β, γ
Protactinium-234	91	5×10^{-14}	1.14 min	β, γ
Uranium-234	92	6×10^{-3}	2.5×10^5 yr	α, γ
Thorium-230	90	2×10^{-3}	8×10^4 yr	α, γ
Radium-226	88	4×10^{-5}	1622 yr	α, γ
Radon-222	86	2×10^{-10}	3.8 days	α
Polonium-218	84	1×10^{-13}	3.1 min	α
Lead-214	82	1×10^{-13}	26.8 min	β, γ
Bismuth-214	83	8×10^{-13}	19.7 min	β, γ
Polonium-214	84	1×10^{-15}	$\sim 10^{-4}$ sec	α
Lead-210	82	5×10^{-7}	22 yr	β, γ
Bismuth-210	83	3×10^{-10}	5 days	β
Polonium-210	84	8×10^{-9}	138 days	α, γ
Lead-206	82		Stable	
Uranium-235 Decay Chain				
Uranium-235	92	0.71	7.1×10^8 yr	α, γ
Thorium-231	90	3×10^{-12}	25.6 hr	β, γ
Protactinium-231	91	3×10^{-5}	3.4×10^4 yr	α, γ
Actinium-227	89	2×10^{-8}	27 yr	α, β, γ
Thorium-227	90	5×10^{-11}	18.6 days	α, γ
Radium-223	88	3×10^{-11}	11.2 days	α, γ
Radon-219	86	1×10^{-16}	3.9 sec	α, γ
Polonium-215	84	6×10^{-20}	$\sim 10^{-3}$ sec	α
Lead-211	82	7×10^{-14}	36.1 min	β, γ
Bismuth-211	83	4×10^{-15}	2.1 min	α, γ
Thallium-207	81	9×10^{-15}	4.8 min	β, α
Lead-207	82		Stable	

* α: alpha particle, a helium nucleus. γ: gamma ray, similar to X-rays. β: beta particle, an electron.

Table 7.4. Principal fission-product radioisotopes in radioactive wastes.

Radioisotope	Atomic number	Half-life	Radiation emitted*
Krypton-85	36	4.4 hr (IT)** → 9.4 yr	β, γ, e⁻ → β, γ
Strontium-89	38	54 days	β
Strontium-90	38	25 yr	β
Zirconium-95	40	65 days	β, γ
Niobium-95	41	90 hr (IT) → 35 days	e⁻ → β, γ
Technetium-99	43	5.9 hr (IT) → 5 × 10⁵ yr	e⁻, γ → β
Ruthenium-103	44	39.8 days	β, γ
Rhodium-103	45	57 min	e⁻
Ruthenium-106	44	1 yr	β
Rhodium-106	45	30 sec	β, γ
Tellurium-129	52	34 days (IT) → 72 min	e⁻, β → β, γ
Iodine-129	53	1.7 × 10⁷ yr	β, γ
Iodine-131	53	8 days	β, γ
Xenon-133	54	2.3 days (IT) → 5.3 days	e⁻, β → β, γ
Cesium-137	55	33 yr	β, γ
Barium-140	56	12.8 days	β, γ
Lanthanum-140	57	40 hr	β, γ
Cerium-141	58	32.5 days	β, γ
Cerium-144	58	590 days	β, γ
Praseodymium-143	59	13.8 days	β, γ
Praseodymium-144	59	17 min	β
Promethium-147	61	2.26 yr	β

*β: beta particle, an electron. γ: gamma ray, similar to X-rays.   e⁻: internal electron conversion.
**IT: isomeric transition, internal.

in geologic formations to reduce future handling and the possibility of leaks to man's environment.

Conversion-to-solid-form methods include use of fluidized beds, heated pots, and radiant-heated spray columns to form solid granules. The addition of glass-forming materials to produce stable, glassy, insoluble products is also under study.

High-level waste storage in natural geologic formations is being investigated. The production of stable solids that are insoluble and noncorrosive reduces the volume of waste to be stored and increases the safety of the storage activities. Hundreds of years will be required for the natural decay of these wastes to harmless levels. Simple burial or disposal in the sea is not a suitable

Table 7.5. Principal activation-product radioisotopes produced by neutron irradiation of nonfuel materials.

Radioisotope	Atomic number	Source and reaction*	Half-life	Radiation emitted**
		Air and Water		
Tritium (H-3)	1	^{2}H (n,γ)	12.3 yr	β, γ
Carbon-14	6	^{14}N (n,p)	5700 yr	β
Nitrogen-16	7	^{16}O (n,p)	7.3 sec	β, γ
Nitrogen-17	7	^{17}O (n,p)	4.1 sec	β
Oxygen-19	8	^{18}O (n,γ)	30 sec	β, γ
Argon-41	18	^{40}A (n,1γ)	1.8 hr	β, γ
		Sodium		
Sodium-24	11	^{23}Na (n,γ)	15 hr	β, γ
Sodium-22	11	^{23}Na (n,2n)	2.6 yr	β^+, γ
Rubidium-86	37	^{85}Rb (n,γ) (impurity in sodium coolant)	19.5 hr (IT)† → 1 min	β, γ → K, γ
		Alloys		
Aluminum-28	13	^{27}Al (n,γ)	2.3 min	β, γ
Chromium-51	24	^{50}Cr (n,γ)	27 days	β^+, K, γ
Manganese-56	25	^{56}Fe (n,p)	2.6 hr	β, γ
Iron-55	26	^{54}Fe (n,γ)	2.9 yr	K
Iron-59	26	^{59}Co (n,p)	45 days	β, γ
Copper-64	29	^{63}Cu (n,γ)	12.8 hr	β, γ, β^+, K
Zinc-65	30	^{64}Zn (n,γ)	250 days	β^+, e$^-$, γ
Tantalum-182	73	^{181}Ta (n,γ)	115 days	β, γ
Tungsten-187	74	^{186}W (n,γ)	24 hr	β, γ
Cobalt-58	27	^{58}Ni (n,p)	71 days	β^+, γ
Cobalt-60	27	^{59}Co (n,γ) (also purposeful irradiation	5.3 yr	β, γ
Phosphorus-32	15	^{31}P (n,γ) (purposeful irradiation)	14.3 days	β

*(n,γ): absorbs neutron, emits gamma. (n,p): absorbs neutron, emits proton.
**β: beta particle, an electron. γ: gamma ray, similar to X-rays. β^+: positively charged electron.
K: orbital electron capture into nucleus. e$^-$: internal electron conversion.
†isomer transition, internal.

answer. Fortunately there are a number of natural geologic formations that might be suitable for permanent storage—they are well removed from man and free of groundwater that might otherwise leach and carry off some of the radioactive material to areas of possible human access.

At present, salt appears to be the best disposal medium. Salt formations are usually dry, impervious to water, and are not associated with usable groundwater sources. The ability of salt to change shape under pressure causes rapid closure of fractures. Salt is sufficiently strong that large cavities formed by mining will not collapse. Salt deposits underlie some 400,000 square miles of the United States. And the estimated volume of high-level waste solids to be produced in the year 2000 would occupy less than 1% of the volume of salt now being mined each year.

Bedrock storage is another possibility. In some cases it may not be necessary to reduce high-level wastes to solid form. Exploratory drilling provides information on the hydrology and geology of rock formations and can verify the compatibility of the rock with the wastes to be stored.

Three principal mechanisms delay the migration of wastes from the bedrock tunnels: a low rate of natural water movement, the impermeability and ion-exchange properties of clay, and the ion-exchange properties of the top layers of earth. Each of these barriers is considered to be an adequate shield in itself. Since they would be encountered in succession, there is an ample factor of safety.

References given at the end of this chapter provide further introductory material. Tables 7.3 through 7.5 provide various information on radioisotopes.

BIOLOGICAL EFFECTS

It has been known for many years that when these radiations strike living things they cause important changes that are often harmful. It is also known that the changes may not be limited to the plant or animal which receives the radiation, but may be passed on to succeeding generations. However, the details of the action, how much radiation will produce a given result, how much can be done to counteract the deleterious effects, are factors that are still under investigation.

There has always been some radiation in the environment. Radium and other radioactive elements in the ground, together with cosmic rays from outer space, produce a natural "background" over all parts of the earth. However, as atomic activity is stepped up throughout the world, the amount of radiation in our surroundings may be substantially increased. This could have profound effects on all forms of life. But there has been disturbingly little information about just what the effects may be .

The problems of radiation fall naturally into two main classes: (1) the effects on human beings; (2) the various ways in which radiation can reach human beings through the environment.

The inheritance mechanism is by far the most sensitive to radiation of any biological system. Any radiation which reaches the reproductive cells causes mutations (changes in the material governing heredity) that are passed on to succeeding generations. Human gene mutations which produce observable effects are believed to be universally harmful. Everyone is subjected to the natural background radiation which causes an unavoidable quantity of so-called spontaneous mutations. Anything that adds radiation to this naturally occurring background rate causes further mutations, and is genetically harmful.

There is no minimum amount of radiation which must be exceeded before mutations occur. Any amount, however small, that reaches the reproductive cells can cause a correspondingly small number of mutations. The more radiation the more mutations. The harm is also cumulative. The genetic damage done by radiation builds up as the radiation is received, and depends on the total accumulated gonad dose received by people from their own conception to the conception of their last child.

So far as individuals are concerned, not all mutant genes or combinations of mutant genes are equally harmful. A few may cause very serious handicaps, many others may produce much smaller harm, or even no apparent damage. From the standpoint of the total and eventual damage to the entire population, every mutation causes roughly the same amount of harm. This is because mutant genes can only disappear when the inheritance line in which they are carried dies out. In cases of severe and obvious damage this may happen in the first generation; in other cases it may require hundreds of generations.

It is difficult to arrive at a figure showing how much genetic harm radiation can do. One measure is the amount of radiation, above the natural background, which would produce as many mutations again as occur spontaneously. It is estimated that this amount is 30 to 80 roentgens.

At present the U.S. population is exposed to radiation from: (a) the natural background, (b) medical and dental X-rays, (c) fall-out from atomic weapons testing, and (d) contaminated industrial waste sites.

Large exposures (say 100 roentgens and up) of the whole body or a large part of it are generally harmful when received in a single dose. Much higher doses may, however, be safely and usefully delivered to limited portions of the body under the controlled conditions of medical treatment. Very little is known about how to treat the pathological effects of radiation or how to protect the body against them in the first place.

One of the obvious effects is a shortening of life. This seems to involve some generalized action. Irradiated individuals may age faster than normally even if they do not develop specific radiation-induced diseases such as leukemia. Exposures small enough to be genetically tolerable have this effect. Furthermore, the permissible exposure levels that have been established for persons working with radiation appear to be within the limits of safety. However, it is not yet known what minimum dose, if any, would be necessary to produce a statistically noticeable reduction of lifespan where very large populations are concerned.

At present, test explosions of atomic weapons are the only significant source of radiation in the general environment, above the natural background. Meteorologists have found no evidence that atomic explosions have changed the weather or climate. Nor do they believe that continued weapons tests, at the same rate and in the same areas as in the past, would have such an effect. Radiation from explosions passes into the atmosphere and much of it eventually returns to the ground as fallout.

Fallout divides into three classes: (1) close-in—material that comes down within a few hundred miles of the explosion and within 10 to 20 hours; (2) intermediate—material that descends in a few weeks after the explosion: (3) delayed—material that remains in the air for months or years.

Close-in fallout from test explosions affects only restricted, uninhabited regions. Intermediate fallout would descend very slowly if it were pulled down only by gravity. It is mostly washed out of the air by rain and snow. It spreads over large parts of the earth, but its effect over a small area may be accentuated if there is heavy precipitation while the radioactive cloud is overhead. Delayed fallout is stored for long periods in the stratosphere. Meteorologists know very little about the interchange of air between the stratosphere and lower layers, so they cannot predict exactly how long the material will stay up, or where it is likely to descend.

It is believed in general that the oceans are not receiving any significant quantities of radioactive material. But they are being considered as a repository for some of the radioactive waste products of atomic power plants. Before this can safely begin on a large scale, much research is needed to determine the mixing rates between various parts of the seas. Materials deposited in some of the deep parts of the ocean may remain there 100 years or more, so that most of their radioactivity would be gone before they reach surface water. On the other hand, material dumped into coastal and other surface waters would directly affect marine life and, within a few years, would contaminate all parts of the world because of the relatively rapid circulation of surface layers. As noted earlier, radioactive tracers can be used to chart ocean and air currents and to study the interrelationships of marine animals.

Radiation from fallout inevitably contaminates the food supply. Radioactive elements in the soil are taken up and concentrated by plants. The plants may be eaten by humans, or by animals, which, in turn, serve as human food. On a global basis the contamination is negligible. But the maximum tolerable level is not known. There is not nearly enough information about the long-term biological effects on man or animals from eating radiation-contaminated food.

Probably the most important potential food contaminant is strontium 90—a radioactive element that concentrates in bone tissue. Already, detectable, yet biologically insignificant, traces of it have turned up in milk supplies thousands of miles from the site of atomic explosions and reactor failures such as the Chernobyl accident in the Soviet Union.

Food from the oceans is also subject to radioactive contamination. Marine

plants and animals extract and concentrate various radioactive elements that
get into sea water. The concentration is cumulative, increasing as it proceeds
up the chain from microscopic plankton to edible fish.

Mutation rates in plants are being artificially accelerated with radiation in
the hope of producing new and superior strains. Thus far, only a few new eco-
nomic varieties have been found, but the method is promising. The use of radiation to sterilize packaged food may have dramatic impact on food technology by reducing the need for refrigeration and extending the shelf-life of many
products.

The major problem in routine disposal is what to do with the wastes resulting from the processing of reactor fuel. The wastes from normal operations of
reactors themselves can be more easily handled. A second major problem is
to anticipate the accidents that will inevitably occur and to set up safety standards which will insure that they do not become catastrophes.

There are the various forms of manmade radiation to reckon with. In indus-
trially advanced countries, by far the most important are medical and dental
X-rays. The average U.S. citizen now receives roughly the same amount of radiation, over his whole body, from X-ray and fluoroscopic examination as
from the natural background.

Another source of radiation is "fallout" from nuclear test explosions. Every
"device" that is set off throws into the air a huge cloud of radioactive particles,
some of which are carried great distances by the wind of the upper air, and set-
tle out gradually over the whole earth.

Every cell in the human body (and in every other living organism) contains
an enormous collection of tiny units called genes. Taken together they substantially determine all the characteristics the individual is born with. Each
person inherits his genes from his parents, who got them from theirs, and so
on. Genes come in pairs, one of each pair inherited from the father and one
from the mother. Every trait is governed by one or more pairs. For example,
there are genes which have to do with hair color, others which control stature
and so on. This is the way the gene pairs are acquired: sperm and egg cells
have single sets of genes rather than pairs. When a sperm and egg unite to
form the first cell of a new embryo, there is a new double set, half coming
from the father and half from the mother. As the embryo develops, this set is
duplicated over and over again through the successive divisions of the cell, so
that every cell in the body of the fully developed infant has an essentially identical set of genes. Just which gene a child inherits from each of his mother's
and father's gene pairs is a matter of chance. Sperm and egg cells are produced
by a splitting of cells in the sex glands or gonads. When the cells split, the
gene pairs divide at random.

From this it is obvious that the two genes in each pair are not necessarily or
even usually identical. Thus, if a child inherits a gene tending to produce red
hair, say, from his father, there is an excellent chance that the corresponding

maternal gene will have some other color tendency. How is the characteristic then determined? What color will the child's hair be? The answer is that in each pair of genes, one almost always has a stronger effect than the other and largely determines the characteristic involved. This is called a dominant gene. The weaker member is called a recessive gene. In the usual case, the recessive's effect is not completely submerged.

Every so often a gene changes (mutates), and a characteristic of the organism is altered. Presumably a changeling gene undergoes some chemical rearrangement, although we do not yet know any of the details of the process. Heat, certain chemicals and especially radiation, will cause mutations, and once changed, the new form of the gene is then passed on as faithfully as the old one was.

Infrequently, a mutation will turn out to be helpful. Evolution has depended on a sequence of rare mutations, each of which produced an organism slightly better equipped than its ancestors to deal with the environment. Plant breeders are continually looking for helpful mutations that will improve crop varieties. But the exceptions merely prove the rule: most mutations are harmful. If mutation takes place in an ordinary body cell, the effect is usually not serious. In any case, the damage is primarily restricted to the individual in whom the change occurred.

The situation is quite different if the gene affected is located in the reproductive cells of the sex glands. Then the damage can be passed on through the sperm or eggs produced by these cells, to the individual's children, thence to their children, and so on.

It is known that radiation causes mutations. Practically all radiation-induced mutations which have effects large enough to be detected are harmful. The change due to a mutated gene seldom is fully expressed in the first generation of offspring of the person who received the radiation, since mutant genes are usually recessive. If from one parent a child gets a mutant gene, but from the other parent inherits a normal gene belonging to that pair, then the normal gene is very likely to be at least partially dominant, so that the normal characteristic will appear. Most of the harm from a radiation-induced mutation might remain unnoticed, for a shorter or longer time, in the genetic constitution of the successive generations of offspring. But the harm would persist, and some of it would be expressed in each generation. On the average, a detrimental mutation, no matter how small its harmful effect, will in the long run tip the scales against some descendant who carries this mutation, causing his premature death or his failure to produce the normal number of offspring. In this way harmful mutations are eventually eliminated from the population. Although many mutations disturb normal development of the embryo, it is not correct to say that all, or even most mutations result in monstrosities or freaks. In fact, the commonest ones are those with the smallest direct effect on any one generation—the slight detrimentals.

It has sometimes been thought that there may be a rate (say, so much per week), at which a person can receive radiation with reasonable safety as regards certain types of direct damage to his own person. But the concept of a safe rate of radiation simply does not make sense if one is concerned with genetic damage to future generations. What counts, from the point of view of genetic damage, is not the rate; it is the total accumulated dose to the reproductive cells of the individual from the beginning of his life up to the time the child is conceived.

In general, the type and severity of pathological effects depend on the amount of radiation received at one time and on the percentage of the total body exposed. Shielding a part of the body reduces the damage in greater proportion than might be expected from the percent of the body mass protected.

Very large single doses (say more than 800 roentgens) which strike all or most of body inevitably cause death. Less than lethal doses produce a variety of effects. The most prominent immediate ones are blood and intestinal disorders; leukemia and cancer are among the chief delayed effects. The skin is very sensitive to radiation. People accidentally exposed to close-in fallout from weapons tests have developed marked external symptoms, including ulcers and loss of hair, although the total radiation they received was not enough to do serious internal damage. Unless the dose is heavy, skin effects are temporary.

REFERENCES

1 Dunlap, C. E. "Delayed Effects of Ionizing Radiation," *Radiology*, 69(1):12–17 (1969).

2 Glasstone, S. and A. Sesonske. *Nuclear Reactor Engineering*. Princeton, NJ:D. Van Nostrand Co., Inc. (1967).

3 Hogerton, J. F. *Atomic Fuel*. Oak Ridge, TN, U.S.:Atomic Energy Commission (Dec., 1967).

4 Mead, R. L. and W. R. Corliss. *Power from Radioisotopes*. Oak Ridge, TN, U.S.:Atomic Energy Commission (Dec., 1966).

5 Miner, W. N. *Plutonium*. Oak Ridge, TN:U.S. Atomic Energy Commission (Aug., 1968).

6 NIOSH Technical Report. "Carcinogenic Properties of Ionizing and Nonionizing Radiation," U.S. Dept. of Health, Education, and Welfare, Pub. No. 78–142 (April, 1978).

7 U.S. Atomic Energy Commission. "Illustrations of Radioisotopes: Definitions and Applications," Technical Information Div., Oak Ridge, TN (1979).

8/ CARCINOGENIC PROPERTIES OF IONIZING RADIATION

IONIZING RADIOCARCINOGENESIS

The earliest reported incident relating cancer to occupational exposure of ionizing radiation was in 1911 when Frieben reported a case of epidermoid carcinoma on the right hand of a radiologist. Dozens of similar cases were reported in the ensuing decades. Within 15 years of Roentgen's discovery of X-rays, Hesse reported at least 94 cases of skin cancer among physicians, X-ray technicians and radium handlers. Multiple squamous cell and basal cell carcinomas of the hand were the most common. Cronkite et al. (1960) have reported over 200 cases of radiation-induced leukemia between the years 1911–1959.

Principle sources of data relating human exposure to ionizing radiation are the military, occupational sources and medical records. These sources of data are largely epidemiological, and few are suitable for establishing quantitative dose-response relationships. Studies of the survivors at Hiroshima and Nagasaki and of patients treated with X-rays, appear to provide the most meaningful statistical data base. Table 8.1 lists various sites of involvement linked to exposure of ionizing radiation. The most common neoplasm is leukemia, followed by cancer of the thyroid, lung, and bone. Ideally, positive confirmation of radiation carcinogenesis consists of discernible dose-response patterns in the epidemiologic data plus confirmation of trends from controlled laboratory experiments on animals.

Ishimaru et al. (1971) have attempted to generate total-dose estimates of the neutron and gamma exposures for the Japanese atomic bombings of 1945. Leukemia data from Hiroshima and Nagasaki differ both in the observed dose-response curves and in the type of leukemia. The population in and around Hiroshima was exposed both to neutron and gamma radiations. Biological effects appear to have been dominated by neutron exposure, and it has been observed by various investigators that the incidence of excess leukemia was

149

Table 8.1. Sites of radiation-induced cancers in man.*

Type of exposure	leukemia	thyroid	breast	lung	stomach	GI organs	skin	bone	head and neck	pancreas	liver	other childhood cancers
Atomic bomb survivors	X	X	X	X	X	X						
Atomic fallout in S. Pacific		X										
Radiologists/dentists	X						X					
Luminous dial painters								X				
Uranium miners and millers				X								
Radium injections and medicines								X	X			
Thorium treatment for spinal tuberculosis								X				

(continued)

Table ε.1. (continued).

Type of exposure	leukemia	thyroid	breast	lung	stomach	GI organs	skin	bone	head and neck	pancreas	liver	other childhood cancers
Thorium dioxide (Thorotrast)	X			X					X		X	
32P-treatment: polycythaemia	X											
X-ray treatment: ankylosing spondylitis	X			X	X	X	X	X		X		
X-ray treatment: enlarged thymus, ring worm, breast, gynecologic disorders, etc.	X	X	X			X	X		X			
X-ray fluoroscopy for tuberculosis			X									
X-ray diagnosis in pregnant women	X											X

*Watson (1975) and from data in Ishimaru et al. (1971), NAS-NRC (1972), Rossi and Kellerer (1974), Upton (1975), Barnett (1976).

roughly proportional to the estimated dose. The ratio of acute lymphocytic to chronic granulocytic leukemia was approximately unity. In Nagasaki, the population was exposed primarily to gamma radiation. Here, the incidence increased with the square of the dose, and the ratio of acute to chronic leukemia was 4:1. There appears to have been a 5-year latency period prior to the onset of excess leukemia in these two cities. The incidence of tumor appearance peaked in 1951 (6 years after exposure), and 25 years later, excess leukemia was still apparent. The tumor sites and number of deaths from other forms of cancer in atom bomb survivors between 1950 and 1970 are summarized in Table 8.2.

A high incidence of leukemia has also been observed among radiologists. A large portion of this data base is associated with early users of X-rays, who were largely unaware of the hazards of ionizing radiation and often accumulated large doses. A higher incidence of leukemia was confined largely to older radiologists, who presumably received much of their exposure in the 1930s–1950s when protective measures were not as widely employed as they are today. Kitabake et al. (1973) review findings for Japanese X-ray workers, which include 28 deaths from leukemia (average latency period, 17.7 years) and 30 deaths from skin cancer (22.2 years average latency). The estimated radiation dose received by these workers is listed in Table 8.3.

Patients having a history of X-ray therapy have also been linked with leukemia and other forms of cancer. It was common practice in the past to treat children with respiratory distress, often diagnosed as having enlarged thymus glands, with therapeutic doses of X-rays. X-ray therapy was given for many other conditions, such as enlarged tonsils, adenoids, and acne. This practice led to 19 cases of thyroid cancer in 2,878 patients (Miller, 1975) and 50 cases of malignant papillary or follicular adenocarcinoma in 1,032 patients (Colman et al., 1975). Colman et al. (1975) report a 19- to 35-year latency period for X-ray induced cancer, and McCohaney and Hayles (1976) note a tumor-induction period of 5–30 years.

Similar observations have been made for children receiving X-ray treatment for ringworm of the scalp and from women receiving repeated X-ray fluoroscopies while undergoing artificial pneumothorax therapy for tuberculosis. In the case of multiple X-ray fluoroscopy, Barnet (1976) reports a four- to eightfold increase in the incidence of breast cancer. There is also a strong correlation between the affected breast and the side of the chest positioned closest to the X-ray source.

Obstetric radiography has also been associated with carcinogenesis. Watson (1975), NAS-NRC (1972), etc. report that diagnostic prenatal X-irradiation increases the risk of leukemia, neuroblastoma, cerebral tumors, Wilms' tumor, and other forms of childhood cancer.

Upton (1975) has linked carcinogenesis to occupational and medical exposure to radionuclides. A study of 770 luminous dial painters who ingested

Table 8.2(A). Deaths from cancer of the digestive organs and peritoneum, 1950*1970.*

Malignant Neoplasm of:	Number of deaths	Percentage
Esophagus	134	5.2
Stomach	1,600	61.7
Small intestine, duodenum	9	0.3
Large intestine (excluding the rectum)	109	4.2
Rectum and rectosigmoid junction	133	5.1
Liver and primary intrahepatic bile ducts	6	0.2
Gallbladder and bile ducts	77	3.0
Pancreas	109	4.2
Peritoneum and retroperitoneal tissue	44	1.7
Unspecified digestive organs	28	1.1
Respiratory and digestive systems (secondary)	344	13.3
	2,593	

*Atomic bomb survivors data from Jablon and Kato (1972).

Table 8.2(B). Deaths from other non-leukemic malignancies, 1950*1970.*

Malignant Neoplasm of:	Number of deaths	Percentage
Tongue	22	4.0
Salivary gland	5	1.0
Lip, gum and mouth	6	1.3
Pharynx	11	2.3
Bone	23	4.8
Connective and other soft tissue	8	1.7
Skin	21	4.4
Ovary, fallopian tube and other female genital organs	51	10.7
Prostate	26	5.5
Testis and other male genital organs	4	0.8
Bladder and other urinary organs	88	18.4
Brain	2	0.4
Thyroid	14	2.9
Other endocrine glands	2	0.4
Ill-defined, secondary and site unspecified	120	25.2
Lymphosarcoma, reticulum cell sarcoma	34	7.1
Hodgkin's disease	21	4.4
Other lymphoid	8	1.7
Multiple myeloma	11	2.3
	477	

*Atomic bomb survivors data from Jablon and Kato (1972).

Table 8.3. Estimated radiation dose received by X-ray workers in Japan.*

Calendar Year	Estimated Dose (R/yr)
1921	900
1922-1926	680
1927-1930	520
1931-1935	350
1936-1940	240
1941-1945	90
1946-1953	9.3
1954-1957	1.55
1958-	0.31

*From Kitabake et al. (1973).

varying amounts of radium (see NAS-NRC, 1972) revealed 51 bone sarcomas and 21 carcinomas of the head and paranasal sinuses. No sarcomas were noted below an accumulated mean bone dose of 900 rads, but the incidence of cancer rose sharply above this value. Clinical and autopsy findings from 42 New Jersey dial painters who survived their initial industrial exposure to 226Ra and 228Ra in the 1910s and 1920s revealed 24 persons with malignancies or blood dyscrasias (Sharpe, 1974).

Saccomanno et al. (1971) report 150 cases of lung cancer among uranium miners who were exposed to cigarette smoke, diesel fumes, and other potential carcinogens. Small-cell and undifferentiated cell carcinomas occurred in 75% of these cases. Archer et al. (1973) reported excess cancers in all groups of uranium miners who exceeded 120 working level month's exposure (WLM) to radionuclides. (1 WLM = 170 hr exposure to 1.3×10^5 MeV of α-particles per liter of air inhaled.)

The use of the tracer thorium dioxide for medical diagnostic purposes (NA-NRC, 1972) has also been linked with carcinogenesis. Da Silva Horta et al. (1972) report a large excess both of hemangioendotheliomas (16/988) and local lesions at the sites of injection. Laboratory animal experiments have also provided information on ionizing radiocarcinogenesis. Some of the observed effects of radionuclides in experimental animals are summarized in Tables 8.4 and 8.5. In general, most animals are like man in that they are susceptible to radiation-induced leukemia. Unlike man, animals exhibit a greater range of tissue sensitivities, even between different strains of the same species. Rats, for instance, are extremely prone to radiation-induced mammary tumors, while burros are resistant to any type of tumor.

The evidence for ionizing radiocarcinogenesis is overwhelming. In many instances, however, the types of tumors induced are not distinct from spontaneous tumors of the same site. Given this fact and the often poorly defined dose-

Table 8.4. Effect of multiple injections or ingestion of ^{89}Sr or ^{90}Sr on experimental animals.*

Animal	Route of Administration	Total Dose (μCi/kg)	Myeloid Leukemia	Myeloid Hypoplasia	Osteosarcoma	Fibrosarcoma	Chondrosarcoma	Angiosarcoma	Reticulum Cell Tumor	Giant Cell Tumor	Lymphoid Leukemia	Reference
Monkey	po	500–1,000	+		+		+					Casarett et al. (1962)
Dog	po	150			+	+	+	+				Pool et al. (1972)
	inj	150	+		+	+		+				Dungworth et al. (1969)
	inj	15	+	+	+	+	+					Finkel et al. (1972)
Swine	po	—	+		+							Finkel et al. (1972)
Rats	po	330–790			+	+		+	+	+	+	Casarett et al. (1962)
	inj	0.1–3.5			+	+		+	+		+	Kuma and Zander (1957)
	inj	4.4					+	+		+	+	Skoryna and Kahn (1959)
												Skoryna et al. (1958)

*For references, see Vaughan (1973). inj = injections; po = ingestion.

Table 8.5. Effects of single injection of ^{89}Sr or ^{90}Sr on experimental animals.[*]

Animal	Route of Administration	Total Dose (µCi/kg)	Myeloid Leukemia	Myeloid Hypoplasia	Osteosarcoma	Fibrosarcoma	Chondrosarcoma	Angiosarcoma	Reticulum Cell Tumor	Giant Cell Tumor	Lymphoid Leukemia	Reference
Dog	iv	63.6–97.9			+			+		+	+	Dougherty and Mays (1969)
Swine	iv	10–500			+	+		+				Finkel et al. (1972)
Rabbits	iv	1,869–6,160			+							Howard et al. (1969)
Rats	iv	50–1,000			+							Vaughan and Williamson (1969)
Mice: CBA	ip	5–500	+									Moskalev et al. (1969)
CBA	iv	20			+			+				Barnes et al. (1970)
	—	200–1,600	+		+	+	+	+				Nilsson (1970)
CF1	iv	41–2,200	+		+	+	+	+				Finkel et al. (1959)

[*] For references, see Vaughan (1973). iv = intravenous; ip = intraperitoneal.

response relationships in human radiocarcinogenesis, direct causality can often be inferred only from known past histories or by epidemiological techniques.

CORRELATING RADIATION DOSE AND RESPONSE INFORMATION

Radiation dose and population response information can be correlated by dose-response relationships. These are important for establishing a causal link between ionizing radiation and carcinogenesis. Once defined, they can help establish adequate safety measures. It is important, then, to know whether radiocarcinogenesis exhibits threshold phenomena. It is equally important to know what physical and host factors are important for each species, tissue, and type of tumor.

There are basically two approaches to developing dose-response relationships. The administrative approach was introduced in 1956 by the International Commission on Radiological Protection (ICRP) (see Watson, 1975). This is a "linear-no threshold" hypothesis that considers any dose of ionizing radiation, no matter how small, to be deleterious. The analysis assumes that the risk of cancer is directly proportional to the dose and is independent of the dose rate and the quality of radiation. The "scientific" approach seeks to uncover the functional form of the dose-response relationship. This approach attempts to define the nature and mechanisms of carcinogenesis and the factors involved in the expression of neoplasms. Of the two approaches, the linear-no threshold hypothesis has the advantages of simplicity and conservatism. At the same time, the linear hypothesis tends to depart from the data, especially at low and high doses. The known dose-response data for lung cancer in American uranium miners is shown in Figure 8.1(A). Similar data for breast cancer in patients receiving diagnostic fluoroscopies are given in Figure 8.1(B). In these examples, the data are adequately described by the linear hypothesis. This is not the case, however, for bone cancer in luminous dial painters or for lung cancer in Newfoundland fluorospar miners as shown in Figure 8.2. In these examples, the data are only poorly approximated by the linear hypothesis. However, the uncertainties are such that the linear hypothesis cannot be eliminated.

In general, animal data tend to be highly nonlinear over wide dose ranges. At low-dose levels, the initiating effects of irradiation dominate and are, to some extent, reversible. At intermediate levels, the initiating effects can be further amplified by other promoting effects. At higher levels, both initiating and promoting effects fail to be fully expressed, due to side effects and excess injury. Data trends for dose-response curves are therefore often sigmoidal,

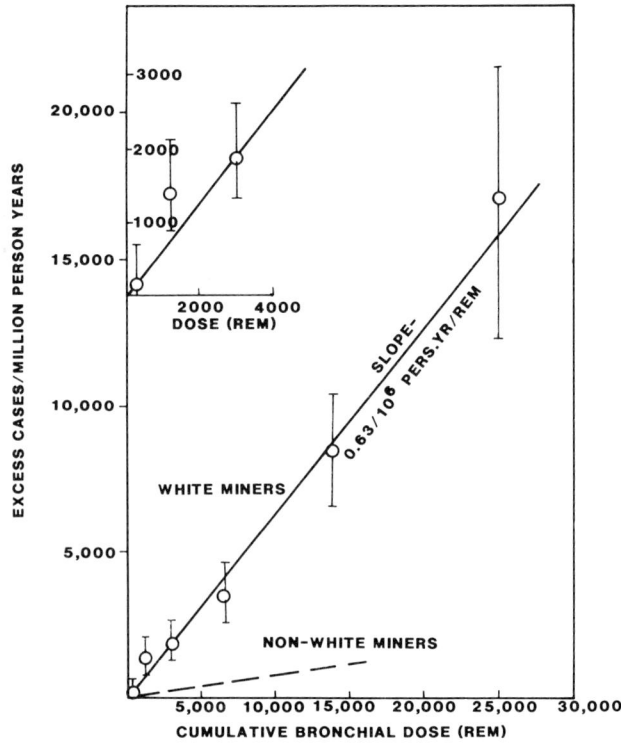

FIGURE 8.1(A). Dose-response data for lung cancer in U.S. uranium miners.

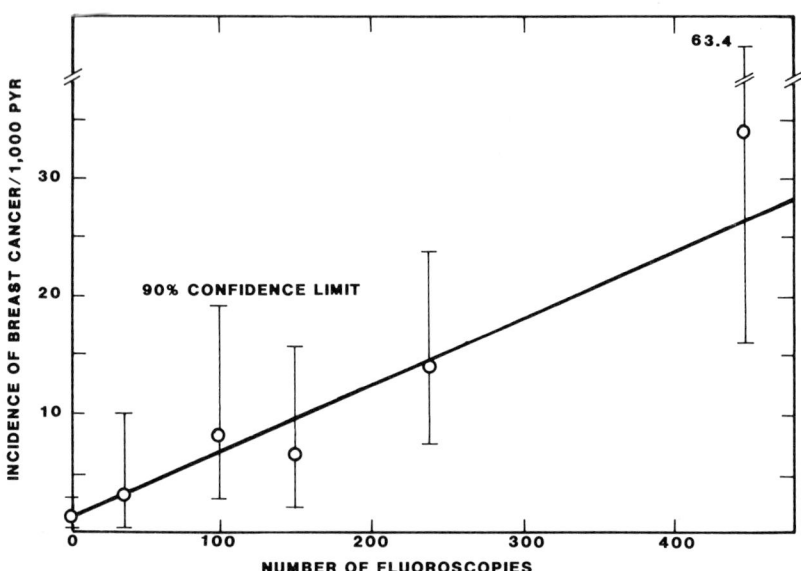

FIGURE 8.1(B). Dose-response data for breast cancer among patients receiving diagnostic fluoroscopies [NAS-NRC (1972)].

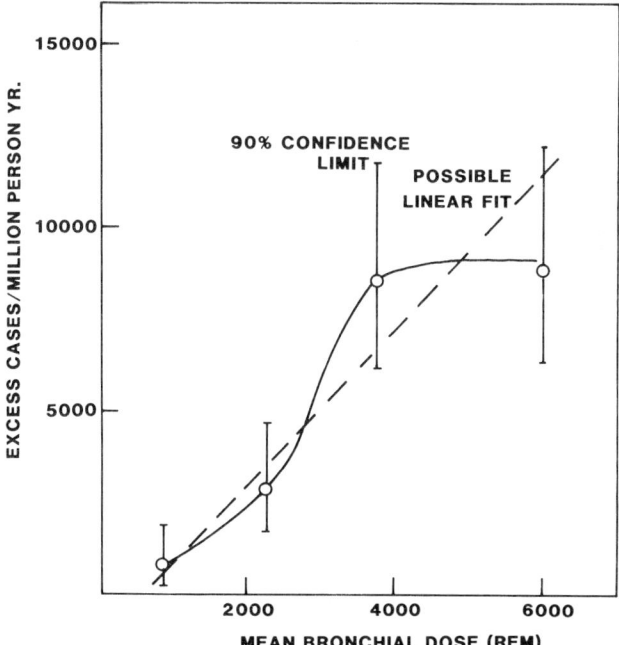

FIGURE 8.2. Dose-response data for lung cancer in Newfoundland fluorospar miners (Villiers et al., 1969).

sometimes exhibiting a threshold and usually dependent on many variables, including the rate and geometry of absorption; the relative biological efficiency (RBE) of the radiation; variations in species, tissue, and cell sensitivity; and differential sensitivity among individuals of the same species. Examples are shown in Figure 8.3.

There are a variety of factors that contribute to large deviation in the dose-response relationships. Fluctuations in energy deposition can be appreciable, especially at low-dose levels. They depend, in part, on the volume in which the energy is to be concentrated, the absorbed dose per hit, and the linear energy transfer (LET) of particles traversing the given volume. At high doses, the number of particles traversing a given volume is large. Statistical fluctuations are relatively small, and the actual energy deposited in a region of given mass (the specific energy) is not likely to differ greatly from the average absorbed dose (see Kellerer and Rossi, 1975; Walburg, 1974). At lower doses, fluctuations become more important, since the number of traversals is lessened. When the average number of traversals per unit volume is less than one, the shape of the observed distribution of energy deposition becomes independent of the dose. A further reduction of dose results in a decrease in the amplitude of the spectrum, with the remainder of the distribution appearing at

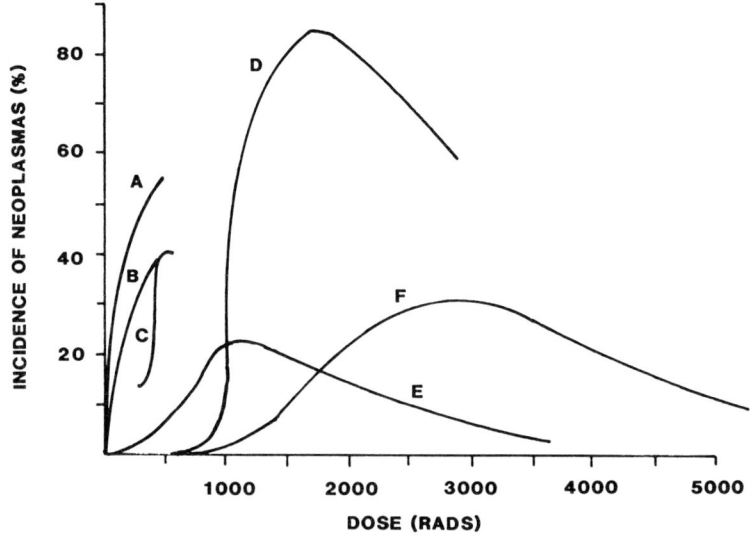

A- MYELOID LEUKEMIA IN MICE BY X-RAYS
B- MAMMARY TUMORS IN 12 MO. OLD RATS BY GAMMA RAYS
C- THYMIC LYMPHOMA IN MICE BY X-RAYS
D- KIDNEY TUMORS IN RATS BY X-RAYS
E- SKIN TUMORS IN RATS BY ALPHA PART.
F- SKIN TUMORS IN RATS BY ELECTRONS (% INCIDENCE X 10)

FIGURE 8.3. Dose-response curves for different types of tumors derived from exposure to external radiation (Walburg, 1974).

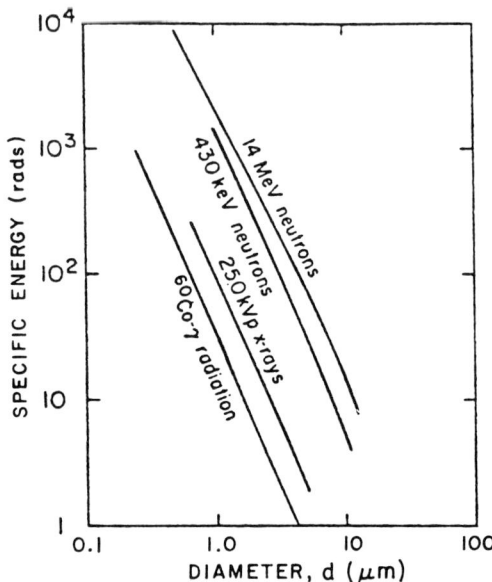

FIGURE 8.4. Mean value of specific energy generated by different ionizing radiations in single events in spherical tissue regions (Kellerer and Rossi, 1975).

the origin. The mean values of the specific energies produced by different types of ionizing radiation in such single events are illustrated in Figure 8.4.

Few studies have been aimed at examining the mutual influence of the total dose, dose rate, and energy accumulation. Nonetheless, some generalizations have been reached (NCRP, 1976). The efficiency of high LET radiation per unit dose is greater than that of low LET radiation and is relatively constant over a wide range of doses and dose rates. The efficiency of low LET radiation decreases with the dose and rate of administration. The RBE of high LET radiation tends to increase with decreasing dose and dose rate. Figures 8.5(A) and (B) show the production of tumors on rats. However, since the total dose, dose rate, and LET may have different effects on the cellular, tissue, organ, and systemic levels, their exact interrelationship may vary with the type of neoplasm, the dose and combination of cells and tissues being irradiated, and the relative susceptibility of individuals in the population.

Data derived from test animals support the conclusion that the form of most oncogenic dose-response curves is sigmoidal. However, for readily induced tumors where host factors are minor, it is difficult to distinguish between linearity and a rapidly rising sigmoidal curve with a very short tail at the low-dose end. Human data support similar inferences, although this data bank is incomplete, especially at the low-dose end where deviations from linearity are expected. It is important to note then that the linear relationships cannot be positively ruled out. The dose-response curve for human populations may be assumed to include a polynomial function comprising a linear-dose term plus one or more power-dose terms. When cell-killing and other high-dose effects

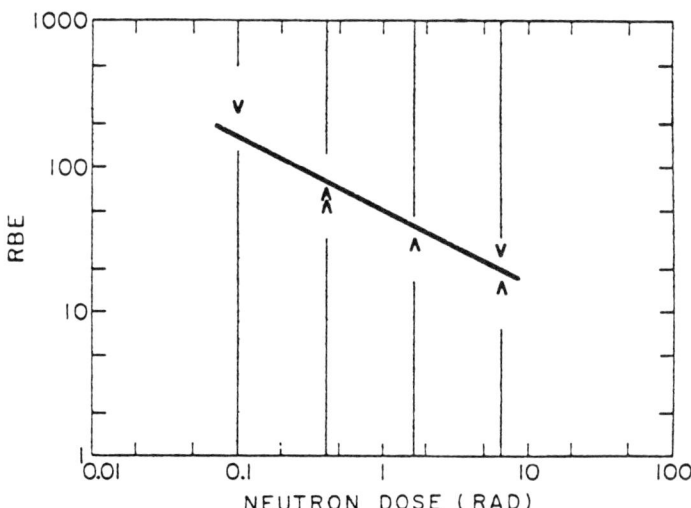

FIGURE 8.5(A). RBE of 430 keV neutrons relative to ionizing radiation for induction of tumors in rats.

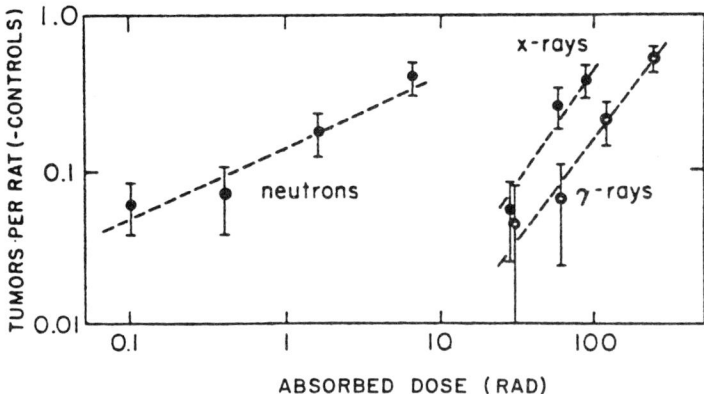

FIGURE 8.5(B). Mean number of tumors per rat minus spontaneous incidence 400 days after exposure to neutrons, X-rays and gamma rays (Kellerer and Rossi, 1975).

are considered, the general form of this curve can be expressed as

$$Y = (C + aD + bD^2)(e^{-\alpha D} + \beta D^2) \tag{8.1}$$

where Y is the expected incidence of cancer, C is the control frequency, D is the absorbed dose, and a, b, α, and β are defined by the system under study (NCRP, 1976).

This type of empirical model may in fact be a gross oversimplification and should never be used for extrapolation purposes. Its importance is in assessing the overall incidence of radiation-induced cancer; however, it cannot be expected to apply equally in every instance. Furthermore, since longer periods are required to accumulate a given dose at low-dose rates, less time remains during the normal life expectancy in which carcinogenic effects may be expressed. Further complexity is introduced (Upton, 1961; Evans, 1974; Jones et al., 1975) by the fact that an increased latency period occurs with a decreasing dose rate, and a similar effect has been seen in animal experiments. An example of this is shown in Figure 8.6. Even apart from the usual age-dependent variability in sensitivity, there is reason to expect a reduction in observed carcinogenicity when exposure is protracted beyond certain limits. The net result is that there may be no single linear term that is strictly invariant over an individual's entire life span, especially at the low-dose rates.

The possibility of protracted exposure at low-dose rates leads to "practical" thresholds. Such thresholds have been observed both in animal studies and in the human data. The data regarding bone tumors in luminous dial painters, for instance, reveal two distinct dose regions. At doses of <900 rads, no bone tumors have been observed; above 1,000–1,200 rads, there is a high incidence of tumors (roughly 30%). These results have been interpreted as implying a latency period exceeding the remaining average life-expectancy of the affected workers. The current state of knowledge is such that, due to the extreme difficulty of proving negative results, few would be willing to state that there is some dose below which there is absolutely no increased probability of developing cancer. Consequently, there appear to be no unique radiation-induced cancers. What is usually observed is an increased cancer incidence seen against a variable "noise level" of spontaneous, but otherwise identical tumors. This variability reduces the statistical certainty with which the natural cancer rate can be utilized as a basis of comparison.

It is important to realize that dose-response relationships are in no way sufficient for projecting potential incidence of cancers. Carcinogenesis involves a variety of complex interactions with many variables and may even depend strongly on ionizing radiation. The dose required for tumor induction may depend not only on irradiation but on a variety of mechanisms, such as the effects of other physical agents, chemical carcinogens, and, at least in certain animal systems, oncogenic viruses. Unfortunately, these mechanisms are poorly understood and less is known about their interactions.

Synergistic effects have been observed in combined therapy with chemotherapeutic agents and therapeutic doses of ionizing radiation. Canellos et al. (1974) report data for cases involving Hodgkin's disease; patients receiving intensive cyclic chemotherapy (MOPP) plus >3,500 R total nodal irradiation had a 10.4% incidence of secondary cancer.

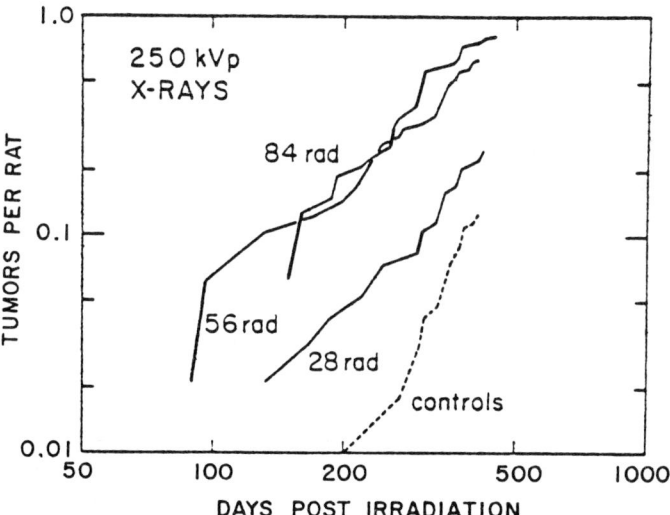

FIGURE 8.6. Mean number of mammary tumors in rats vs. time after exposure to different doses of 250 kV$_p$ X-rays (Kellerer and Rossi, 1975).

The three carcinogenic agents most widely studied for synergism are radiation, chemical carcinogens, and viruses. Animal studies have focused largely on the chemical carcinogens. Particular emphasis has been placed on their effect on irradiated skin, liver, and bronchial tissues; some of the early studies are cited in the reference section of this chapter. Many of the results of these studies appear to be contradictory, due to wide methodological variations, many of which are not yet comparable. Despite this, these data do suggest several qualitative trends. Most early experimenters concluded that for some sites, such as the lung, ionizing radiation actually suppressed the development of chemically-induced cancer. (See Heston et al. and Duplan.) Other investigators, however, have noted positive synergistic effects at lower doses (Lindop and Rotblat, 1966), concluding that irradiation inhibits bronchial carcinogenesis in rats only when the dose exceeds a certain threshold. This concept, if correct, would go far in explaining the apparent contradictions in the literature. Implicit in such a concept is the view that chemical carcinogens promote not only the carcinogenic effects, but also the cell killing and other high-dose effects of ionizing radiation. When the combined effect of radiation and chemical toxicity exceeds a certain level, the cell killing effects will dominate; below this level, neoplastic transformations are more probably. At present, however, the existence of synergistic thresholds remains unproven.

The literature addressing the interaction of ionizing radiation and oncogenic viruses is sparse. Frazier (1975) reports the presence of reverse transcriptase, which is indicative of retravirus or oncornavirus infection, in radiation-induced canine osteosarcomas. Others (see Rubin and Bakemeir, 1974) have suggested that radiation-induced leukemia is mediated through the activation of latent RNA viruses. Experiments with normally resistant X/GF mice (Goldfeder, 1976) suggest that X-rays and other immuno-suppressive agents activate endogenous viral genomes in target mammary gland cells. A positive contributory role for radiation has yet to be established in human cancers; however, there is growing evidence to suggest the presence of oncogenic viruses. At present, there have been few systematic studies of these tissues.

PREDICTING OCCUPATIONAL HAZARDS

In general, dose-response, synergistic, and mechanistic data are largely incomplete. This makes any attempt of the prediction of excess cancers in occupational settings (i.e., at low dose levels and rates) extremely difficult. Predictions can only be based on the dose-response curve for the specific effect(s) in question. When only fragmentary dose-response data are available, extrapolation to the low-dose range requires a theory regarding the underlying mechanisms, the dose rate, the statistical distribution of individual thresholds within the population at risk, and the appropriate latent period(s), as influenced by both the dose and the dose rate. Many regulative bodies, therefore, have adopted the conservative linear no-threshold hypothesis.

With this approach, any dose of ionizing radiation is considered deleterious, no matter how small it may be. The risk of cancer is considered to be directly proportional to the dose and to be independent of the dose rate and quality of radiation. Thus, if, as NAS-NRC (1972) estimates, about 1% of the spontaneous cancers in the U.S. or 3,000–4,000 deaths per year, are caused by the average background radiation of 0.1 rem/year, the cancer incidence at higher levels of exposure is predicted to be a linear multiple of this rate. Radiation protection from this point of view is based on balancing risks against the benefits accrued. The advantage is that it allows the relation of incidence to a simple mean-accumulated dose characterizing the exposure of different groups under nonuniform conditions (NAS-NRC 1972) to be estimated.

An example of a typical risk estimate based on the linear hypothesis is shown in Table 8.6 (NAS-NRC, 1972). This table summarizes the NAS-NRC estimated risk of leukemia based on data from Japanese survivors and patients receiving therapeutic doses of X-rays. In both groups, the risk of leukemogenesis appears to be greater for those under 10 years of age at time or irradiation. The absolute risk is estimated to be 1–3 excess cases per 10^6 persons per year per rem for adults and 1–8 cases for children. The corresponding percent increases in relative risk (incidence in the exposed group divided by incidence in unexposed persons) are 2–4 and 7–11 per rem in adults and children, respectively.

The linear approach to risk estimates is conservative and tends to overestimate the carcinogenic effects of both low and high doses of ionizing radiation. NAS-NRC (1972) reports no detectable excess of leukemia in populations exposed at dose rates comparable with present occupational exposure limits. UNSCEAR (1974) reports that the actual incidence of radiation-induced leukemia approximates that of all other radiation cancers, and ICRP (1965) estimates the incidence of other radiation-induced cancers to be 2–3 times the rate of leukemogenesis. Jacobi (1976) estimates an actual mean cancer risk of not greater than 0.001% per rad at doses < 50 rads, and Frigerio et al. (1973) note that the number of deaths from cancer actually decreases with small increments in background radiation. Thus, the true risk of cancer from occupational exposure to ionizing radiation may not significantly exceed the spontaneous rate. If, on the other hand, one accepts the NAS-NRC (1972) estimate of 3,000–4,000 deaths per year due to background radiation (15–20 deaths per 10^6 persons per year), the linear hypothesis predicts a major increase in the absolute incidence of cancer due to occupational exposure. The risk for those receiving the maximum allowed cumulative whole-body dose (5 times the person's age beyond 18 years of age, an average of 5 rem/yr) would be 750–1,000 deaths per 10^6 persons per year of exposure above the spontaneous rate of 1,500–2,000 per 10^6 per year. The absolute risk of occupa-

Table 8.6. Risk estimates for leukemia.*

Population	Irradiated as Adults (>10 yr)		Irradiated as Children (1–10 yr)	
	Absolute Risk (cases/10^6/yr/rem)	Relative Risk (% increase in rate ppm)	Absolute Risk (cases/10^6/yr/rem)	Relative Risk (% increase in rate/rem)
Atom Bombs Survivors (1945)				
Hiroshima and Nagasaki	0.76–1.8	3.7	1.1–3.3	6.5
Hiroshima Alone	0.86–2.6	4.3	—	—
Nagasaki Alone	0.09–0.98	2.3	—	—
Spondylitics (1935–1954)	1.0–1.7	2.9	—	—
Menorrhagia Patients	0.5–2.4	2.7	—	—
Thymus Patients (1926–1957)	—	—	1.3–5.6	8.1
Tinea Capitis Patients (1940–1949)	—	—	0.9–7.7	11

*Summarized from data reviewed in NAS-NRC (1972).

tionally induced death from cancer prior to age 65 for radiation workers exposed to the maximum cumulative safety level since age 18 would be 35,000–47,000 per 10^6 persons or 3.5–4.7%. Allowing for a 10-year induction period, these figures are reduced to 27,500–37,000 occupationally-related deaths from cancer per 10^6 workers so exposed. Although compatible with the uncertainties in the known data, these figures appear to be high and serve to emphasize an over-conservatism by their use of so simple a method of risk estimation.

A primary goal in radiation protection is the prediction of risks incurred at various exposure levels. Unfortunately, there is no reliable means of doing this. Linear extrapolation is unreliable at low levels of exposure and ignores many of the factors which contribute. A polynomial model seems technically more correct, but the epidemiological data is inadequate for a proper statistical regression. Accurate estimates of the margin of safety afforded by current practices are therefore not possible.

REFERENCES

Albert, R. E. and B. Altshuler. "Considerations Relating to the Formulations of Limits for Unavoidable Population Exposures to Environmental Carcinogens." in *Radionuclide Carcinogenesis*, Sanders et al., eds. U.S. Atomic Energy Commission, pp. 233–233 (1973).

Alexander, P. *Progr. Exp. Tumor Res.*, 10:22–71 (1968).

Archer, V. E., G. Saccomanno and J. H. Jones. "Frequency of Different Histologic Types of Bronchogenic Carcinoma as Related to Radiation Exposure," *Cancer*, 34: 2056–2060 (1974).

Archer, V. E., J. K. Wagoner and F. E. Lundin. "Lung Cancer Among Uranium Miners in the United States," *Health Phys.*, 25:351–371 (1973).

Barnett, M. H. "The Biological Effects of Ionizing Radiation: An Overview," HEW Publication (FDA), 77–8004 (1976).

Becker, F. F. "Etiology: Chemical and Physical Carcinogenesis," in *Cancer. A Comprehensive Treatise, Vol. 1*. Plenum Press (1975).

BEIR Report (1972): See NAS-NRC (1972).

Brent, R. "Radiations and Other Physical Agents," in *Handbook on Teratology, Vol. 1*. Wilson & Fraser, eds. Plenum Press, pp. 153–223 (1977).

Brown, J. M. "Lineariety vs. Non-Linearity of Dose-Response for Radiation Carcinogenesis," *Health Phys.*, 31:231–245 (1976).

Brown, D. G., D. F. Johnson and F. H. Cross. "Late Effects Observed in Burros Surviving External Whole-Body Gamma Irradiation," *Radiation Res.*, 25:574 (1965).

Burns, F. J., I. P. Sinclair, R. E. Albert and M. Vanderlaan. "Tumor Induction and Hair Follicle Damage for Different Electron Penetrations in Rat Skin," *Radiat. Res.*, 67:474–481 (1976).

Canellos, G. P., V. T. DeVita, J. C. Arseneau and R. C. Johnson. "Carcinogenesis by

Cancer Chemotherapeutic Agents: Second Malignancies Complicating Hodgkin's Disease in Remission," *Res. Results Cancer Res.*, 49:180–114 (1974).

Cole, L. J. and P. C. Nowell. pp. 393–419 in *Immunity and Tolerance in Oncogenesis*. Severi, ed. Div. Cancer Res., Perugia (1970).

Colman, M., L. R. Simpson, L. K. Patterson and J. Ovadia. "Irradiation of the Neck During Childhood: an Assessment of Factors which Influence the Development of Thyroid Tumors in Persons at Risk" (Abstract), *Radiat. Res.*, 62:599 (1975).

Conrad, R. A. "Thyroid Nodules and Leukemia in a Marshallese Population Exposed to Fallout 20 Years Ago" (Abstract), *Radiat. Res.*, 59:313 (1974).

Conrad, R. A. "Twenty-Year Review of Medical Findings in a Marshallese Population Accidentally Exposed to Radioactive Fallout," NTIS Report, BNL 50424 (1975).

Cronkite, E. P., W. Moloney and V. P. Bond. "Radiation Leukemogenesis, An Analysis of the Problem," *Am. J. Med.*, 5:673–682 (1960).

Da Silve Horta, J., J. D. Abbatt, L. Cayolla da Motta and M. H. Tavares. "Leukemia, Malignancies and Other Late Effects Following Administration of Thorotrast," *Z. Krebsforsch.*, 77:202–216 (1972).

Da Silva Horta, J., L. Cayolla da Motta and M. H. Tavares. "Thorium Dioxide Effects in Man. Epidemiological, Clinical and Pathological Studies. (Experience in Portugal.)", *Environ. Res.*, 8:131–159 (1974).

Diamond, E. L., H. Schmerler and A. M. Lilienfeld. "The Relationship of Intra-Uterine Radiation to Subsequent Mortality and Development of Leukemia in Children. A Prospective Study," *Amer. J. Epidemiol.*, 97:283–313 (1973).

Elkind, M. M. and J. L. Redpath. "Molecular and Cellular Biology of Radiation Lethality," to be published in *Cancer: A Comprehensive Treatise, Vol. 6*. F. F. Becker, ed. Plenum Press (in press).

Evans, R. D. "The Effect of Skeletally Deposited Alpha-Ray Emitters in Man," *Brit. J. Radio.*, 39:881 (1966).

Evans, R. D. "Radium in Man," *Health Phys.*, 27:497–510 (1974).

Evans, R. D., A. T. Keane, R. J. Kolenkow, W. R. Neal and M. M. Shanahan. "Radiogenic Tumors in the Radium and Mesothorium Cases Studied at MIT," in *Delayed Effects of Bone-Seeking Radionuclides*. Mays et al., eds. Salt Lake City: University of Utah Press, pp. 157–194 (1969).

Evans, R. D., A. T. Keane and M. M. Shanahan. "Radiogenic Effects in Man of Long-Term Skeletal Alpha-Irradiation," in *Radiobiology of Plutonium*. Stover and Jee, eds. The J. W. Press, pp. 431–468 (1972).

Finkel, M. P. and B. O. Biskis. *Progr. Exp. Tumor Res.*, 10:72–111 (1968).

Finkel, M. P., C. A. Reilly and B. O. Biskis. "Pathogenesis of Radiation and Virus-Induced Bone Tumors," *VIth Intl. Symp. of the Gesellschaft zur Bekampfung der Krebskranheiten Nordrhein-Westfalen e.V.* Heidelberg: Springer-Verlag, pp. 92–103 (1976).

Frazier, M. E., J. F. Park, W. S. S. Jee and G. Taylor. "Investigation of the Role of On-cornavirus in Radiation-Induced Osteosarcomas," NTIS Report Conf. 75–1043–6 (1975).

FRC (1960): *Report No. 1: Background Material for the Development of Radiation Protection Standards*. Government Printing Office.

Frigerio, N. A., K. F. Eckerman and R. S. Stowe. "The Argonne Radiological Impact Program (ARIP). Part I. Carcinogenic Hazard from Low-Level Low-Rate Radiation," NTIS Report W-31-109-Eng-38 (1973).

Furth, J. and E. Lorenz. "Carcinogenesis by Ionizing Radiations," in *Radiation Biology, Vol. 1*. Hollaender, ed. McGraw-Hill, pp. 1145–1201 (1954).

Gitlin, J. N. "Preliminary Dose Estimates from the U.S. Public Health Service 1970 X-Ray Exposure Study," paper presented at the *49th Annual Meeting of the American College of Radiology*, Miami Beach, FL, April 6, 1972 (1972).

Goldfeder, A. "Induction of Intracytoplasmic Type B Virions in a Mouse Mammary Tumor," (Abstract), *Proc. Am. Assoc. Cancer Res.*, 17:12 (1976).

Harley, N. H. and B. Bennett. Personal communication referenced by Warren (1974).

Harley, N. H., R. E. Albert, R. E. Shore and B. S. Pasternack. "Follow-Up Study of Patients Treated by X-Ray Epilation for Tinea Capitis. Estimation of the Dose to the Thyroid and Pituitary Glands and Other Structures of the Head and Neck," *Phys. Med. Biol.*, 21:631–642 (1976).

Hirashima, K. and T. Kumatori. "The Increasing Susceptibility of Hematopoietic Stem-Cells to Friend Leukemia Virus after X-Irradiations" (Abstract), *Radiat. Res.*, 59:91 (1974).

ICRP. "The Evaluation of Risks from Radiation," *Health Phys.*, 12:239–302 (1966).

Ihle, J. N., J. C. Lee and H. G. Hanna. "Characterization of Natural Antibodies in Mice to Endogenous Leukemia Virus," in *Biology of Radiation Carcinogenesis*. Yuhas et al., eds. Raven Press, pp. 261–273 (1976).

Illmensee, K. and B. Mintz. "Totipotency and Normal Differentiation of Single Teratocarcinoma Cells Cloned by Injection into Blastocysts," *Proc. Nat. Acad. Sci.*, 73:549–553 (1976).

Ishimaru, T., T. Hoshino, M. Ichimaru, H. Okada, T. Tomiyasu and T. Tsuchimoto. "Leukemia in Atomic Bomb Survivors, Hiroshima and Nagasaki, 1 October 1950–30 September 1966," *Radiation Res.*, 45:216–233 (1971).

Jablon, S. and H. Kato. "Studies of the Mortality of A-Bomb Survivors, 5. Radiation Dose and Mortality, 1950–1970," *Radiation Res.*, 50:649–698 (1972).

Jacobi, W. "Relation Between Radiation Dose and Somatic Radiation Risk," *Atomwirt Atomtech*, 19:278–283 (1976).

Jones, H. B., A. Grendon and M. R. White. "The Time Factor in Dose-Effect Relations," in *Proc. Int. Sympos. on Biol. Effects of Low-Level Radiation Pertinent to Protection of Man and His Environment*, IAEA, Chicago, Nov. 3.7, 1975. (1975).

Kellerer, A. M. and H. H. Rossi. "Biophysical Aspects of Radiation Carcinogenesis," pp. 405–439 in Becker (1975).

Key, M. "Radiobiology of Endocrine Organs," in *Hormones and Cancer*. Charyulu and Sudarsanam, eds. Stratton Intercontinental Med. Book Corp., pp. 21–33 (1976).

Kitabake, T., T. Watanabe and S. Koga. "Radiation Cancer in Japanese Radiological Workers," *Strahlentherapie*, 146:599–606 (1973).

Klein, J. C. "The Use of In Vitro Methods for the Study of X-Ray Induced Transformation," in *Biology of Radiation Carcinogenesis*. Yuhas et al., eds. Raven Press, pp. 301–307 (1976).

Kondrateva, A. F. "Combined Action of Ionizing Radiation and Chemical Carcinogens (A Review of Experimental Findings," *Voprosy Onkologii*, 15:109–118 (Rus.) and NTIS report: NIH-72-28 (1969).

Lacassagne, A. "Les cancers produits par les rayonnements corpusculaires; mecanisme presumable de la cancerisation par les rayons," in *Actualities Scientifieques et Industrielles, No. 981*. Paris:Hermann et Cie (1945).

Lacassagne, A. "Les cancers produits par les rayonnements electromagnetiques," in *Actualites Scientifiques et Industrielles, No. 975*. Paris:Hermann et Cie (1945 b).

Levy, R. L., M. H. Barrington, R. A. Lerner and F. J. Dixon. "On the Mechanism of Infectivity of a Murine Leukemia Virus in Adult Mice," in *Biology of Radiation Carcinogenesis*. Yuhas et al., eds. Raven Press, pp. 275–286 (1976).

Lieberman, M., H. S. Kaplan and A. Decleve. "Anomalous Viral Expression in Radiogenic Lymphomas of C57BL/Ka Mice," in *Biology of Radiation Carcinogenesis*. Yuhas et al., eds. Raven Press, pp. 237–244 (1976).

Lindop, P. J. and J. Rotblat. *Nature*, 210:1392–1393 (1966).

Miller, R. W. "Radiation," in *Cancer Epidemiology and Prevention*. D. Schottenfeld. Springfield, IL: C. C. Thomas, pp. 93–101 (1975).

Mintz, B. and K. Illmensee. "Normal Genetically Mosaic Mice Produced from Malignant Teratocarcinoma Cells," *Proc. Nat. Acad. Sci.*, 72:3585–3589 (1975).

Misiti-Dorello, P., G. Cancelliere, G. DeMartino and M. Quintiliana. "Modification of Gamma-Ray Sensitivity of Bacterial Membrane by Pre-Exposure to Light," *Radiat. Environ. Biophys.*, 13:19–26 (1976).

Modan, B., H. Mart, D. Baidatz, R. Steinitz and S. G. Levin. "Radiation-Induced Head and Neck Tumors," *Lancet*, 7852:277–279 (1974).

Mole, R. H. "Antenatal Irradiation and Childhood Cancer: Causation or Coincidence?" *Br. J. Cancer*, 30:199–208 (1974).

Munoz, N. "Prenatal Exposure and Carcinogenesis," *Tumori*, 62:157–162 (1967).

McConahey, W. M. and A. B. Hayles. "Thyroid Neoplasia and Radiation to the Head, Neck and Upper Thorax of the Young," *J. Pediatr.*, 89:169–170 (1976).

McLean, A. S. *Atom*, 294 (Sept., 1963).

McRee, D. I. "Environmental Aspects of Microwave Radiation," *Environ. Health Perspectives*, 2:41–53 (1972).

NAS-NRC. *Report of the Advisory Committee on the Biological Effects of Ionizing Radiations (BEIR Report): The Effects on Populations of Exposure to Low Levels of Ionizing Radiation* (1972; reprinted July, 1974).

NCRP. "Review of the Current State of Radiation Protection Philosophy," NCRP Report No. 43, Washington, DC (1975).

NCRP. "Influence of Dose Rate and LET on Dose-Effect Relationships: Implications for Estimation of Risks of Low-Level Irradiation," report prepared by NCRP Scientific Committee 40, July, 1976 (1976).

NIOSH Technical Report 78–142. "Carcinogenic Properties of Ionizing and Nonionizing Radiation," Vol. 3—Ionizing Radiation (April, 1978).

Oleinick, N. L. and R. C. Rustad. "Interrelationships Between Ionizing Radiation, Protein Synthesis and the Physiological Expressions of Radiation Damage," *Adv. in Radiation Biol.*, 6:107–160 (1976).

Rasmussen, N. C. 'Reactor Safety Study: An Assessment of Accident Risks in U.S. Commercial Nuclear Power Plants," WASH-1400, NUREG-75/014 (1975).

Rossi, H. H. and A. M. Kellerer. "The Validity of Risk Estimates of Leukemia Incidence Based on Japanese Data," *Radiation Res.*, 58:131–140 (1974).

Rubin, P. and R. F. Bakemeier. *Clinical Oncology for Medical Students and Physicians*. 4th edition. American Cancer Society (1974).

Saccomanno, G. V., E. Archer, O. Auerback, M. Kuschner, R. P. Sauders and M. G. Klein. "Histologic Types of Lung Cancer Among Uranium Miners," *Cancer*, 27:515 (1971).

Sacher, G. A. "Dose, Dose Rate, Radiation Quality and Host Factors for Radiation-Induced Life Shortening," in *Aging, Carcinogenesis and Radiation Biology—The Role of Nucleic Acid Addition Reactions*. Smith, ed. Plenum, pp. 493–517 (1976).

Seltser, R. and P. Sartwell. "The Influence of Occupational Exposure to Radiation on the Mortality of American Radiologists and Other Medical Specialists," *Amer. J. Epid.*, 81:2–22 (1965).

Severi, L., ed. *Immunity and Tolerance in Oncogenesis*. 2 vols. Perugia, Italy:Divs. Cancer Res. (1970).

Sharpe, W. D. "Chronic Radium Intoxication: Clinical and Autopsy Findings in Long-Term New Jersey Survivors," *Environ. Res.*, 8:243–383 (1974).

Shellabarger, C. J., V. P. Bond, E. P. Cronkite and G. E. Aponte. "Relationship of Dose of Total-Body ^{60}Cobalt Radiation to Incidence of Mammary Neoplasia in Female Rats," *Radiation-Induced Cancer*, pp. 161–172, Vienna:IAEA (1969).

Stone, R. S. "Common Sense in Radiation Protection Applied to Clinical Practice," *Am. J. Roentgenol. Radium Ther. Nuclear Med.*, 78:993–999 (1957).

Stone, R. S. "Maximum Permissible Exposure Standards," in *Protection in Diagnostic Radiology*. Rutgers Univ. Press (1959).

Storer, J. B. "Radiation Carcinogenesis," pp. 453–483 in Becker (1975).

Takeichi, N. "Induction of Salivary Gland Tumors in Rats Following X-Ray Irradiation. II. Development of Salivary Gland Tumors in Long Term Experimentation," *Hiroshima J. Med. Sci.*, 23:391–411 (1975).

Tennant, R. W., J. A. Otten, J. M. Quarles, W. K. Yang and A. Brown. "Cellular Factors that Regulate Radiation Activation and Restriction of Mouse Leukemia Viruses," in *Biology of Radiation Carcinogenesis*. Yuhas et al., eds. Raven Press, pp. 227–236 (1976).

UNSCEAR. *Report of the United Nations Scientific Committee on the Effects of Atomic Radiation*, General Assembly, Official Records, Nineteenth Session, Supp. 14 (A/5814), New York (1964).

UNSCEAR. *Report of the United Nations Scientific Committee on the Effects of Atomic Radiation*, General Assembly, Official Records, Twenty-first Session, Supp. 14 (A/6314), New York (1966).

UNSCEAR. *Ionizing Radiation: Levels and Effects*, United Nations, New York (1972).

Upton, A. C. "The Dose-Response Relation in Radiation-Induced Cancer," *Cancer Res.*, 21:717 (1961).

Upton, A. C. "Radiation Carcinogenesis," in *Methods in Cancer Research, Vol. IV*. H. Busch, ed. Academic Press, pp. 53–82 (1968).

Upton, A. C. "Physical Carcinogenesis: Radiation—History and Sources," pp. 387–404 in Becker (1975).

U.S.P.H.S. *Population Dose from X-Rays, U.S., 1964*, PHS Pub. No. 2001 (1969).

van Bekkum, D. W. "Mechanism of Radiation Carcinogenesis" (Abstract), *Radiat. Res.*, 59:8 (1974).

Vaughan, J. *The Effects of Irradiation on the Skeleton*. Oxford:Clarendon Press (1973).

Vaughan, J. "Some Illustrative Systems of Radiation Carcinogenesis: (2) Bone Seeking Isotopes and Neoplasis," in *Scientific Foundations of Oncology*. Symington and Carter, eds. Year Book Medical Publishers, pp. 456–466 (1976).

Walburg, H. E. "Experimental Radiation Carcinogenesis," *Adv. Radiat. Biol.*, 4:209–254 (1974).

Walters, R. A. and M. D. Enger. "Effects of Ionizing Radiation on Nucleic Acid Synthesis in Mammalian Cells," *Adv. in Radiat. Biol.*, 6:1–48 (1976).

Warren, S. "Effects of Occupational and Environmental Exposures to Ionizing Radiation," NTIS Report No. C00-3017-17 (1974).

Watson, G. M. "Effects of Ionizing Radiation on Man," pp. 1–27 in AAEC/IP1. Coogee, Australia:Australian Atomic Energy Commission (1975).

Yuhas, M. J. "Dose-Response Curves and Their Modification by Specific Mechanisms," in *Biology of Radiation Carcinogenesis*. Yuhas et al., eds. Raven Press, pp. 51–61 (1976).

9 / RADIATION PROTECTION *AND* CONTROL

GENERAL CONTROL PRECAUTIONS

Where suitable precautions are taken and controls implemented, the handling of radiation sources is a safe operation. Proper precautions can only be implemented after a thorough evaluation of the potential hazards to the environment is made. External hazards can be minimized by reducing all external radiation levels to values which are as low as practicable. This can be accomplished by a variety of methods. One example is to ensure that the minimum radiation output, in the case of a radiation source such as an X-ray machine, or the minimum quantity of radionuclide is used for any specified operation. Other methods of minimizing hazards include maintaining the maximum possible distance compatible with effective working methods between the radiation source and the worker; limiting the time spent in the vicinity of the source to the minimum necessary; employing proper shield between the source and the worker and, where necessary, using additional shielding to ensure that other persons in the vicinity or in adjoining areas are appropriately protected from radiation exposure.

Precautions against internal hazards include conducting all operations with open sources in enclosures such as fume cupboards and glove boxes; ensuring that good housekeeping habits are maintained in all areas where open sources are handled; and applying extreme care to ensure that any radioactive spill is confined to well-defined areas by using special appliances such as trays and, in the event of an uncontrolled spill, taking immediate steps to prevent further spread of any air, surface, effluent, personnel or other contamination.

With all sources of radiation, conducting periodic surveys aimed at ensuring that the degree of air, surface, effluent and personnel contamination is well within acceptable limits, is mandatory.

Sealed sources of radiation such as those found in a hospital which has facilities for brachytherapy (cancer therapy with sealed sources) often require very

173

special handling methods. This application involves threading of radium or cobalt needles, each of a few millicuries strength. The safe execution of such an operation involves first ensuring that the needles are safely stored with sufficient shielding when not in use and that appropriately shielded transport containers are provided for transporting them to their place of use. Next, one should remove one needle at a time for threading while the remaining needles are held in their secondary shielded storage container. Shielding the needle (except the eyelet) with an appropriate lead container during the threading operation to minimize radiation exposures during threading is also required. The actual implantation operation should also be conducted in the minimum time possible with maximal shielding facilities and suitable remote handling devices. Patients who have radiation source implants in them should themselves be treated as radiation sources and hence, should be suitably segregated. The sources in their bodies should be provided with adequate external shielding to ensure that other patients or medical or paramedical personnel are not unduly exposed to radiation.

Similar problems arise in the handling of industrial radiography sources. For example, a camera with a suitably collimated source is employed in industrial radiography, adequate steps should be taken to segregate the operation and to provide appropriate shielding for all persons involved. Similar steps of a more stringent nature will have to be taken in the case of panoramic exposures. Area monitoring instruments and, where necessary, radiation alarms should be provided so that any accidental situation that could arise could be immediately detected and adequate remedial measures taken.

In cases of open sources of radionuclides, even more stringent precautions are needed. Once a radionuclide is released in an uncontrolled manner to the environment, it is very difficult to decontaminate or to contain the spread of contamination. This is particularly true of ingestion and inhalation hazards, since radionuclides, once they gain entry into the body, cannot easily be eliminated. Thus, it is important to effectively contain radionuclides at all stages of handling. General methods of controlling and containing radiation sources are described below.

METHODS OF SHIELDING

Shielding involves the use of suitable radiation absorbing material placed between the source of radiation and the personnel exposed in order to reduce the intensity of the radiation to acceptable levels. This reduction in intensity is known as attenuation.

The absorbing material used and the thickness required to attenuate the radiations to acceptable levels depend upon the type of radiation, its energy, the flux and the dimensions of the source. The amount of shielding material required may be calculated with reasonable accuracy for some cases. The adequacy of shielding must be verified by measurements of the intensity of the radiation with suitable instruments. The absorbing material should be installed as close to the source as possible to obtain maximum economy; for, although the thickness required is the same, the area and hence the total volume of the absorber will be greatly reduced.

The specific shielding material depends in part on the type of radiation. For example, alpha particles lose energy rapidly in passage through matter and hence, do not penetrate very far. For the energy range of alpha particles usually encountered, a fraction of a millimetre of any ordinary material is sufficient to absorb them. Thin rubber, perspex, stout paper or cardboard will absorb them.

In contrast, beta particles do not lose energy so rapidly and are more penetrating than alpha particles. Their range in air varies from about 1 m at 0.5 MeV to about 10 m at 3 MeV, but in comparatively dense material their range is very much reduced. Materials composed mostly of elements of low atomic number such as perspex, aluminum and thick rubber are most appropriate for the absorption of beta particles. For example, 1/4 inch of perspex will absorb all beta particles up to 1 MeV and 1 inch of perspex will absorb all beta particles up to an energy of 4 MeV. Hence, it is advisable to use materials of low atomic number for beta radiation shielding since, generally speaking, only a very small pecentage of the beta radiation energy will give rise to bremsstrahlung from these materials. With high-energy beta particles from large sources, the bremsstrahlung contribution may become significant and it may be necessary to provide additional shielding of high atomic weight material (such as lead) to attenuate the bremsstrahlung radiation.

Neutrons are uncharged particles and are therefore capable of considerable penetration in matter. The use of shielding to attenuate a neutron beam should be directed towards reducing the energy of the neutrons to levels at which they can be easily absorbed. A reduction of the energy of neutrons is best accomplished by collisions with atoms of light elements (e.g., hydrogen). For neutrons of energies above 1 MeV the use of heavier elements is also effective. Neutron interaction in these materials produces inelastic collisions from which the neutrons are ejected with reduced energy and gamma photons are also emitted. Additional light elements must be used, however, to reduce the neutron energy to below that at which neutron capture becomes possible. The absorption of neutrons gives rise to secondary-particle and/or gamma-ray emission.

Water and paraffin wax are easily handled hydrogenous compounds which are effective for reducing the energy of fast neutrons. For example, 10 inches of paraffin wax will attenuate 1 MeV fast neutrons by about a factor of 10. A thin sheet of cadmium (about 1 mm in thickness) is adequate for the absorption of neutrons of thermal energy. Since absorption can be followed by gamma ray emission, it may be necessary to provide additional shielding of lead or other similar material to attenuate the gamma radiation.

The attenuation of X- and gamma-rays in an absorbing material is the result of a combination of the photoelectric effect, the Compton effect and pair production. The photoelectric effect is the predominant type of interaction at low energies, the Compton effect at medium energies and pair production at very high energies. Because of the increasing cross sections for interaction in materials of high atomic number, the most suitable materials for this type of shielding are lead and iron. In the medium energy range (0.5–0.75 MeV), the density of the material is more important than the atomic number, but in the higher and lower energy ranges, materials of higher atomic number are more effective.

The scattering of X- and gamma-rays in passing through the absorbing material involves consideration of shielding calculations for narrow beam and broad beam geometrical conditions.

Consider a narrow collimated beam of gamma radiation which might be obtained from a small source such as is used in gamma radiography. Photons scattered in the shielding material placed in the collimated beam are removed from the emergent beam as shown in Figure 9.1(A). The total attenuation is

$$I = I_0 e^{-\mu x} \tag{9.1}$$

where

I = the intensity (flux) of the radiation beam emerging from the shielding material, in photons/cm^2 · s
I_0 = the intensity (flux) of the beam incident on the shielding material
μ = the linear aborption coefficient for the shielding material referred to a thickness of 1 cm
x = the thickness of the shielding material, in cm

The half-value thickness of a material is related to its linear absorption coefficient by the relation:

$$\text{HVT} = \frac{0.693}{\mu} \tag{9.2}$$

Similarly,

$$\text{TVT} = \frac{2.303}{\mu} \tag{9.3}$$

The reciprocal of the linear absorption coefficient has dimensions of length

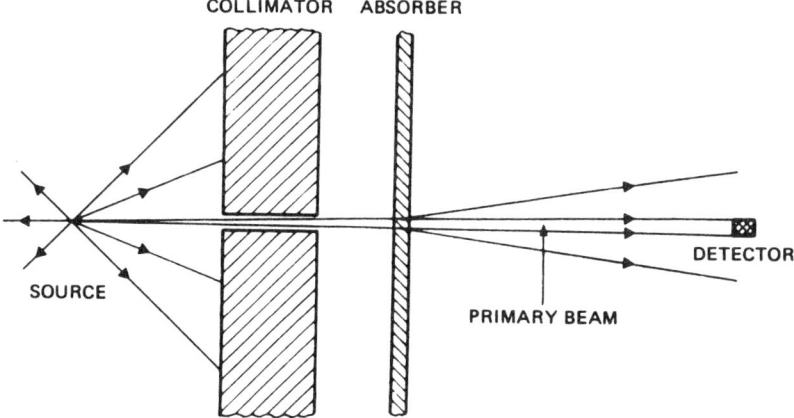

FIGURE 9.1(A). Narrow beam condition.

[known as the <u>mean free path</u> (MFP)]. A thickness of 1 MFP will produce an attenuation to $1/e$, 2 MFP an attenuation to $1/e^2$, etc.

The linear absorption coefficient varies with the energy of the photon, the atomic number of the material, and the density of the material. The term <u>mass absorption coefficient</u>, which is equal to the linear absorption coefficient divided by the density, is also used. Mass absorption coefficient values reflect the relative efficiencies of different shielding materials for the same weight of the various materials.

Broad beam conditions are more frequently encountered in radiation shielding problems. When broad parallel beams or divergent beams of radiation pass through attenuating material, some scattered radiation re-enters the emergent beam as shown in Figure 9.1(B). Thus, the attenuation no longer follows an exponential process, as it is reduced by an amount known as the "build-up factor" for any particular source-shield arrangement.

Protection of occupied areas adjacent to X- and gamma-ray installations may be achieved by shields or walls which absorb radiation. As the cost of such shielding can prove to be an important consideration, judicious siting of these installations at a distance from occupied areas will be advantageous. For example, an isolated basement which is surrounded on all sides by the walls of its excavation and which therefore only needs appropriate shielding material on the roof may provide an economical site.

Protective barriers can be of two types—the primary protective barrier, which is a barrier sufficient to attenuate the useful beam to the permissible levels, and the secondary protective barrier, which is a barrier sufficient to attenuate the scattered and/or leakage radiation to the permissible levels.

Overall requirements of shielding can be minimized if the useful beam is controlled such that it is used only in certain specified directions. The appara-

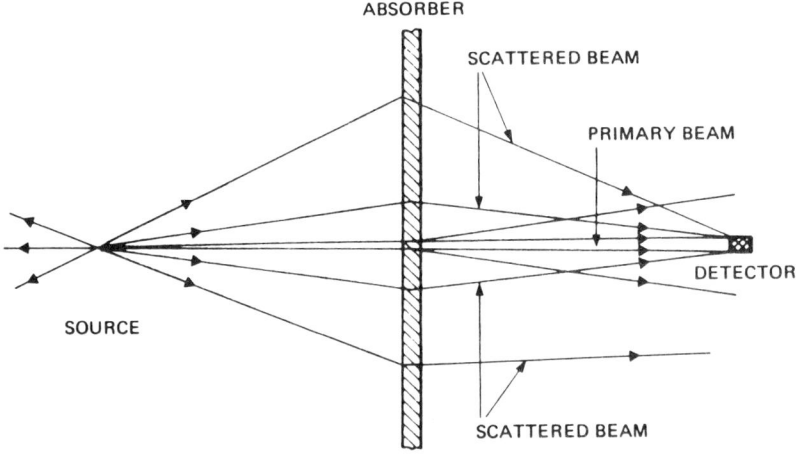

FIGURE 9.1(B). Broad beam condition.

tus should be provided with suitable electrical or mechanical interlocks to ensure that the useful beam can only be used in the chosen directions. Where these directions may be chosen so as to avoid the walls of adjacent occupied areas, further economies can be effected in shielding requirements.

The protection afforded by shielding must be spatially continuous. Hence, special attention should be given to joints, bolts or openings to ensure that suitable overlaps of the absorbing material are provided.

Figures 9.2–9.10 provide specific data on radiation attenuation through various materials.

To ensure that adequate protection is afforded, the shielding thickness for an X-ray machine should be calculated for the maximum rated voltage and current of that particular machine. Similarly, for gamma-ray installations, shielding calculations should be based on the maximum strength of the gamma source proposed to be employed. For both X- and gamma-ray installations, the shortest distances between the source and occupied areas should be considered. Other factors to be taken into account are the work load, the usage factor and the occupancy factor.

The work load is related to the maximum expected extent of use of an X-ray or gamma-ray source in a specified period of time. For X-ray equipment operating below 4 MV, the work load is usually expressed in milliampere minutes per week. For gamma beam therapy sources, and for X-ray equipment operating at 4 MV or above, the work load is usually stated in terms of the weekly exposure of the useful beam at one meter from the source and is expressed in roentgens per week at one meter.

The use factor indicates the fraction of the work load during which the radiation beam under consideration is directed at a particular barrier.

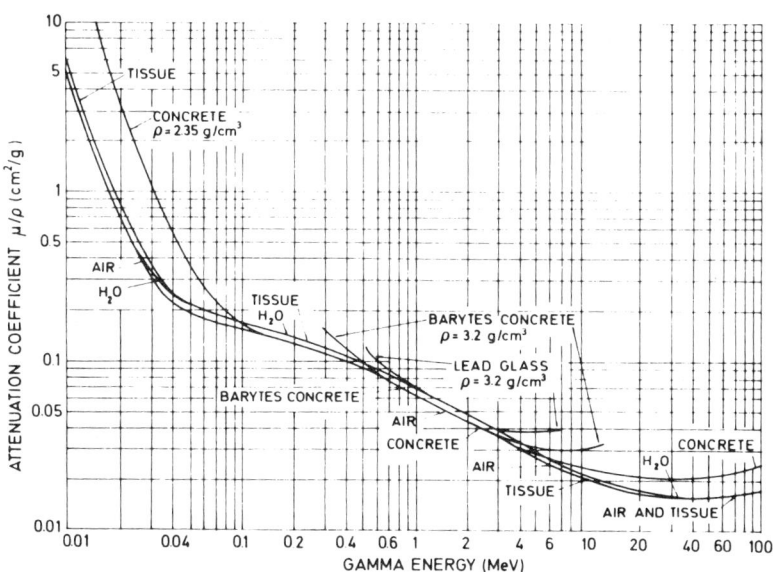

FIGURE 9.2. Mass attenuation coefficient for different materials.

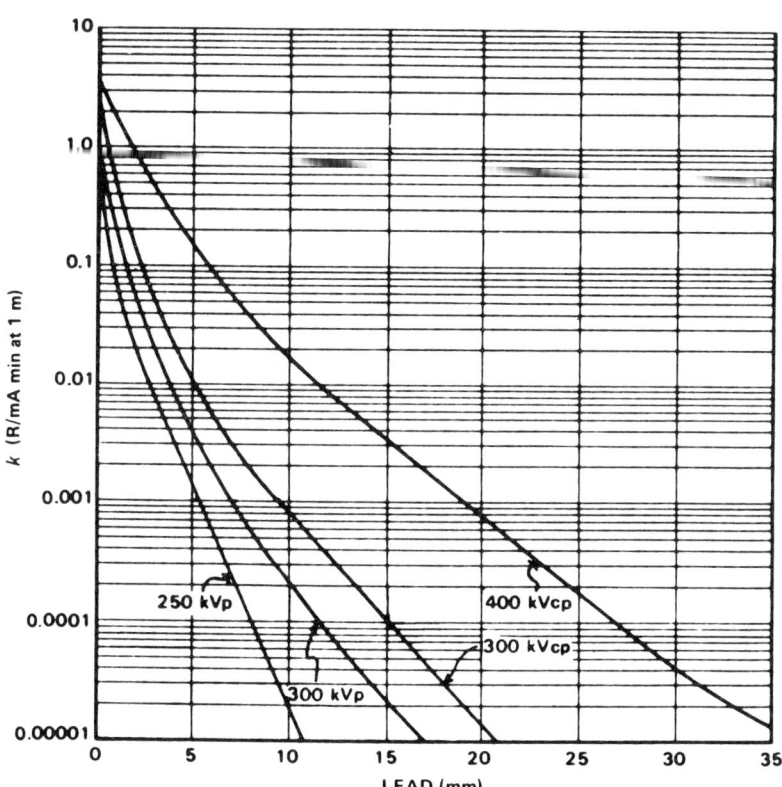

FIGURE 9.3. Attenuation in lead of X-rays (potentials 250–400 kv).

179

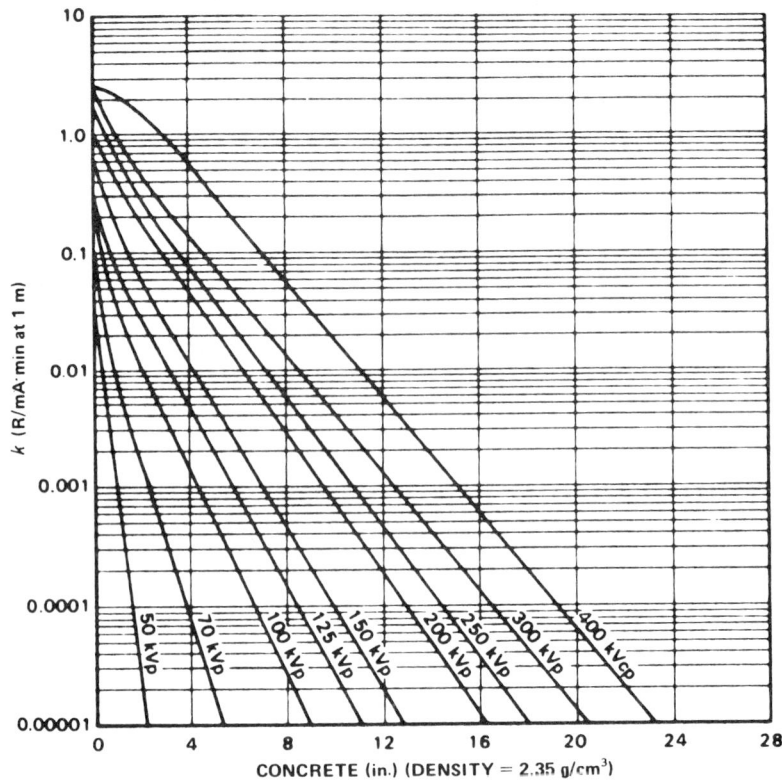

FIGURE 9.4. Attenuation in concrete of X-rays. Concrete density 2.4 g/cm³ and X-ray potential 50–400 kV.

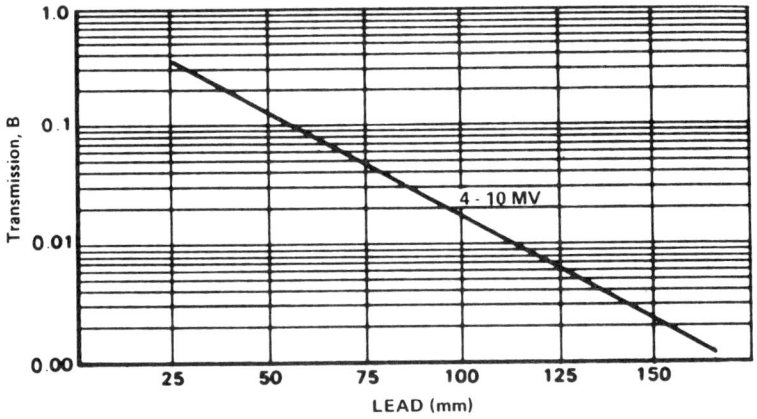

FIGURE 9.5. Attenuation in lead of X-rays (4–10 MV).

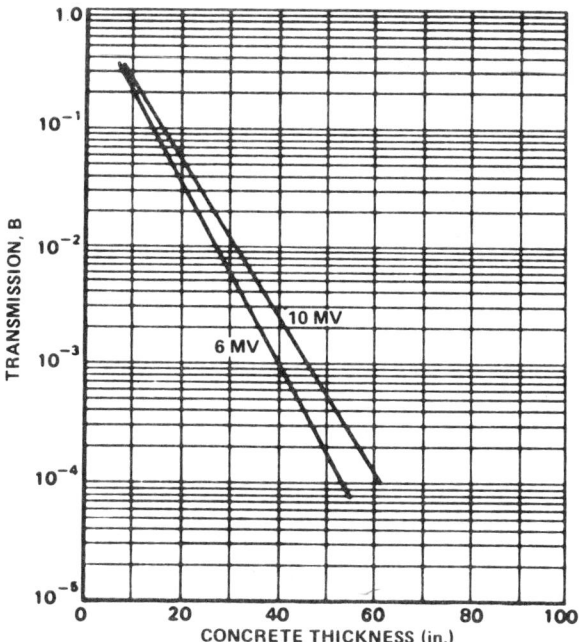

FIGURE 9.6. Attenuation in concrete of X-rays (concrete density = 2.4 g/cm³; radiation potential = 6 and 60 Mv).

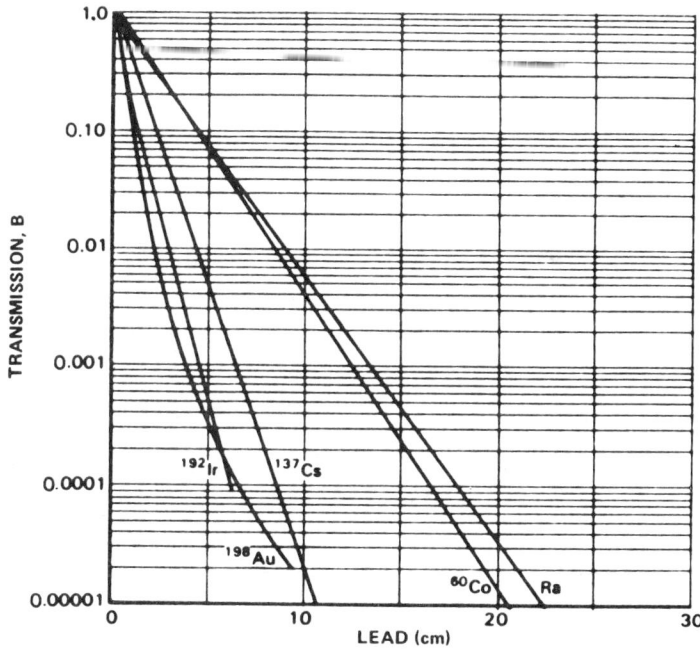

FIGURE 9.7. Shows transmission through lead of gamma rays generated from various radionuclides.

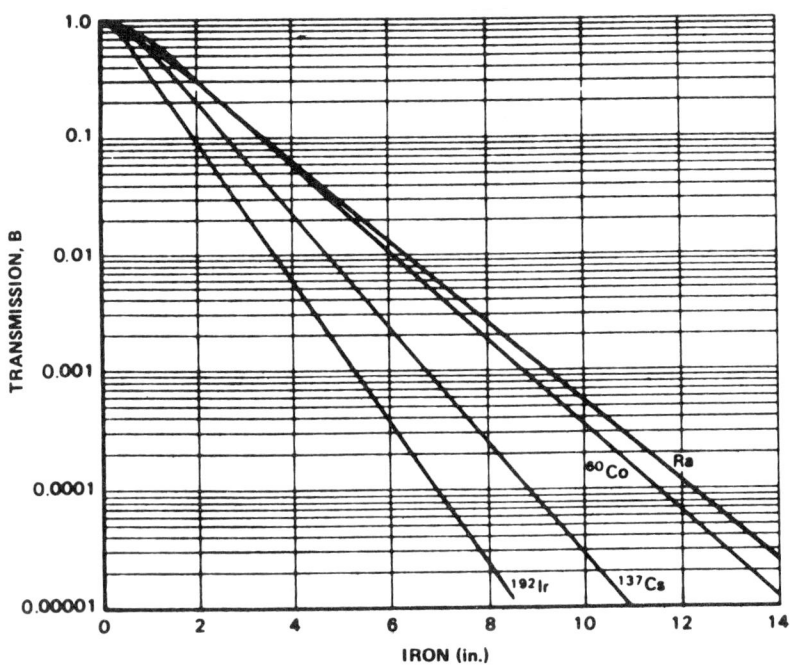

FIGURE 9.8. Transmission through iron of gamma rays from various radionuclides.

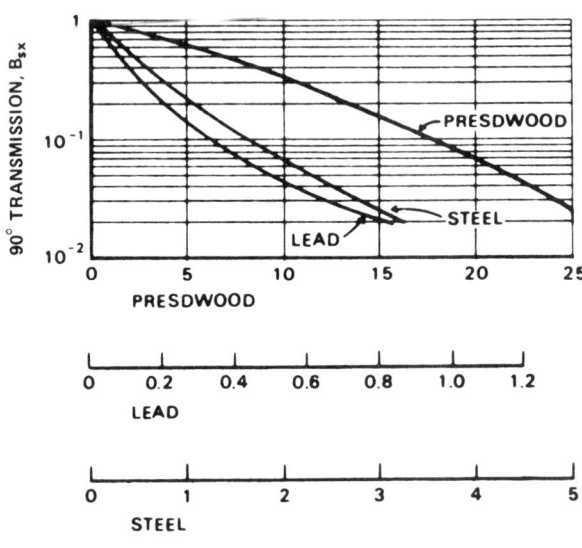

FIGURE 9.9(A). Transmission through lead, Presswood and steel of 6-MW radiation.

182

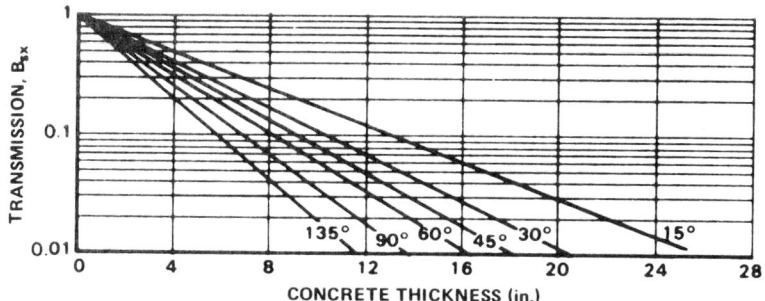

FIGURE 9.9(B). Broad beam attenuation in concrete for 6 MV X-rays scattered at different angles from a Presswood phantom.

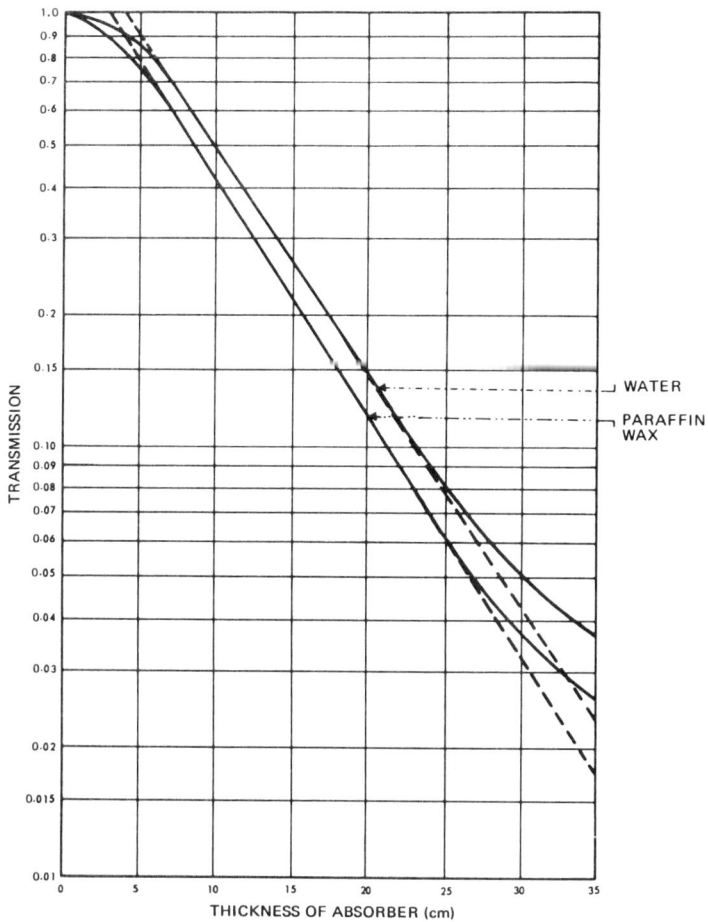

FIGURE 9.10. Shows fast neutron transmission in paraffin wax and water for Po-Be neutron source.

The occupancy factor is the factor by which the work load should be multiplied to correct for the degree of occupancy of the area in question while the source is operating.

For any occupied area in the vicinity of an X- or gamma-ray source, the shielding provided is such that the integral dose which takes into account the work load, the use factor and the occupancy factor, (1) will not exceed 100 mR per week if occupational radiation workers are involved, and (2) will not exceed 10 mR per week if persons who are not occupationally exposed to radiation are involved. The weekly maxima of 100 mR per week and 10 mR per week have been arrived at from the annual maxima of 5 R and 0.5 R respectively for occupational workers and members of the public.

PROTECTIVE CLOTHING

Protective clothing for radioactive work may be considered under two headings—routine and emergency. The former generally includes laboratory coats, aprons, gloves, footwear, and respirators. Laboratory coats consisting of conventional white cotton drill or nylon coats of proper size extending below the knees are suitable for clean areas.

Overalls or boiler suits are comprised of one-piece cotton drill garments designed to cover the body completely except for the head and neck, wrists and hands, and feet and ankles. The fastenings for these garments are usually at the front. They are extremely useful, since they protect all the clothing worn below. Where the processes involve working at benches with liquids, an apron of suitable impervious materials such as PVC, polyethylene or neoprene will be found useful in preventing the clothing below from becoming contaminated by corrosive liquids or dust.

Rubber gloves are often used for general laboratory work. Surgical gloves are adequate for most operations. Where it is necessary to handle beta-active material directly with the hands, rubber gloves of a heavier type or leather gloves may be used to reduce the beta-radiation dose to the hands.

Footwear should preferably be rubber-soled to prevent the uptake of contamination and to facilitate cleaning. It is recommended that the pattern of the rubber sole should not be too deeply indented. The upper part of the shoes should be well waxed to resist the absorption of contaminated solutions, etc.

For work in areas of low or medium level airborne activity, a full-face respirator with an efficient filter provides adequate protection. The filter used must be reliable; suggested types are the resin wool and charcoal or the highly efficient paper filters which are commercially available. Care must be taken to ensure that these respirators fit properly and do not allow air to be taken in from the sides of the face-piece. For work in areas of very low activity, a half-face respirator may be used.

Emergency protective clothing usually consists of garments so designed that they cover the individual completely, except for the head and neck and the hands and feet, with a layer of impervious material. As a protection for the head, and to allow a respirator or breathing apparatus to be worn at the same time, hoods of similar material are used. Finally, complete exclusion from contamination is obtained by wearing rubber gauntlet gloves pulled well over the cuffs of the protective suit and rubber boots with the bottom of the impervious suit trousers brought over them.

Pressurized clothing is comprised of a suit made of impervious material which completely encloses the individual. Such a suit effectively isolates the individual from surface and airborne contamination. Compressed air supplied to the suit enables normal breathing during operations. The compressed-air line delivers its air immediately in front of the face, since this arrangment provides plenty of air for breathing and at the same time helps to reduce the misting of the transparent head-piece. Complete protection of the hands and wrists is afforded by wearing rubber gloves, which are securely taped to the suit to prevent the ingress of any contamination. Rubber boots are usually worn with the suit.

Where a compressed-air supply is not available for use in the pressurized suits described above, breathing sets may be employed. These comprise a well-fitting face-piece in which suitable goggles are inserted. This face-piece is attached to a cylinder or cylinders of compressed air carried in a harness on the wearer's back. Regulating valves attached to the cylinders enable the wearer to control his air supply. This breathing set may be used in conjunction with the fully protective impervious clothing. So dressed, an individual can enter a contaminated atmosphere under emergency or maintenance conditions.

Routine protective clothing should be cleaned regularly to avoid the build-up and fixation of contamination. For similar reasons, the clothing should not be left in storage for long periods before cleaning, as experience indicates that such storage results in the contamination being fixed and renders decontamination increasingly difficult. Another measure taken in such clothing to avoid build-up and fixation of contamination is to provide no pockets or belts and to minimize the folds.

MONITORING AND DECONTAMINATION PRACTICES

Radiation monitoring of the working environment and surrounding areas is an essential part of any effective radiation protection program to ensure that neither the operating personnel nor the general population receives radiation doses in excess of permissible limits. The type and extent of the environmental monitoring program for a particular radiation installation will largely depend on the individual circumstances. A continuing assessment of the radiation situation in the working environment is necessary to ensure safe working conditions. This procedure is an essential adjunct to the personnel monitoring system which measures the radiation doses received by operating personnel.

In cases where open sources are to be handled, preliminary surveys should be conducted, where appropriate, to determine background radionuclide concentrations in the environment. Such studies on environmental radionuclide concentrations should be continued to ensure that during operations radionuclides are being released to the environment in a controlled manner. For example, in a hospital, this could mean periodic monitoring of the effluents discharged from the site.

Depending upon the individual circumstances, the working environment may be monitored by the use of instruments such as installed radiation monitors, portable survey meters, air monitors and contamination monitors. Off-site monitoring is normally carried out by methods of routine air and effluent sampling.

Installed radiation monitors are invariably fitted with visual and aural indicators so that when a predetermined level of radiation is reached, this fact is immediately brought to the attention of the operator. In areas of potential criticality accidents, area monitors consisting of TLD or glass dosimeters, chemical dosimeters, track detectors and neutron activation detectors are installed, In case of an accident these monitors are immediately retrieved and dose estimates made. In large installations, remote-monitoring devices for measuring radiation levels and airborne and surface contamination levels are installed with devices for continuously relaying information to a central recording facility.

Routine area monitoring both inside any radiation working area and also in its immediate environment is an important aspect. Portable survey meters which mainly use ion chambers, GM counters and scintillators are used in these operations. Where there is potential for high radiation intensities either as part of normal planned operations or as a result of abnormal or accident situations, fish-pole probe-type survey instruments with high ranges should be used.

Routine monitoring of the air is a valuable means of ensuring the effectiveness of safety precautions taken to prevent undue release of radioactivity either to a working area or to the wider environment. Monitoring is particularly essential in those laboratories where significant quantities of unsealed radioactive substances are handled.

Airborne radioactive contamination may consist of fine particles of radionuclides or their compounds in a pure form or in association with dust particles. When contaminated air is drawn through a special filter paper, most of the particulate matter is retained on the filter paper. The radioactivity collected on the filter paper provides a measure of the contamination in the air.

Sampling devices may be permanently installed or portable. Installed air samplers may be useful in laboratories handling significant quantities of radioactive substances and also in stacks which exhaust air from fume cupboards and other similar enclosures used for handling open sources. Portable air samplers are used for spot sampling in areas where airborne contamination is suspected.

The release of radioactive wastes to the environment is based on the principles of "concentrate and contain," or "dilute and disperse." In the case of the latter, it is important that the capacity of the environment to absorb the amount of radioactivity released should be carefully pre-estimated, and the release should be based on such estimates. To ensure that the releases are indeed taking place as planned, effluents may have to be monitored either periodically or on a continuing basis depending on the volume and complexity of the operation involving radioactive materials.

Off-site monitoring systems may include devices such as continuous air samplers or water samplers which are remotely installed and from which the readings are continuously relayed to a control center. While in some cases installed monitors may be essential, in other cases portable or mobile sampling units would suffice. In addition to air and water sampling, periodic sampling of vegetation and soil and other biological indicators in the vicinity may be useful for monitoring the release of radioactivity to the environment.

Decontamination is the process of removal of radioactive contamination from the skin or from surfaces such as the walls or floor of working areas. Decontamination can prove to be an expensive operation, in terms of both time and money. Hence, the main aim in the design and operation of any working place for radioactive substances should be to reduce the possibilities of contamination to the absolute minimum.

In decontaminating the skin, it would be ideal to remove the entire contamination. This may not always be possible because the drastic measures which may be necessary in certain cases could result in such damage to the skin that the radioactive material could gain entry into the body and so give rise to an internal hazard. In such cases, it should be considered satisfactory to reduce the levels of contamination to within permissible limits. Similar considerations would apply to the decontamination of surfaces such as walls, floors, and table tops, and of contaminated equipment.

It is important to establish maximum permissible levels of contamination for the skin and for surfaces in controlled and uncontrolled areas. Some representative values for these levels have been provided in the IAEA Manual, Safe Handling of Radionuclides, Safety Series No. 1, 1973 Edition.

The fundamental principles which are applicable to all decontamination procedures are:

- Wet decontamination methods should always be used in preference to dry.

- Mild decontamination methods should be tried before resorting to treatment which can damage the surfaces involved.
- Precautions must always be taken to prevent the further spread of contamination during decontamination operations.
- Where possible, contamination involving short-lived activities should be isolated and segregated to allow natural decay to take its course.

Soap and water is the first requirement for removing contamination from the hands and other exposed areas of the skin of exposed personnel. The soap chosen should be mild so that it will not produce skin damage after frequent use.

For hands, a soft-bristle nail brush should be provided for use in conjunction with soap and water over the entire surface of the hands and the wrists. Particular attention should be given to the nails, to the ridges between the fingers and to the edges of the hand. Frequent rinsing is essential during the entire operation.

For the face, copious amounts of water and soap should be used. Isolated areas of high contamination should be carefully scrubbed. All personnel should be instructed to keep the eyes and the mouth closed during treatment and to rinse the face frequently with copious amounts of water. While using towels or other materials suitable for drying, rubbing should be avoided. All cases of face contamination should be referred to the medical officer.

In case of contaminated small open wounds, cuts, punctures, etc., the wound should be immediately washed, bleeding should be encouraged if necessary, and the medical officer should be consulted.

Special decontamination apparatus, such as vacuum cleaners fitted with special filters, may be used to remove loose contamination from working areas. No attempt should be made to brush or dust it off, though in the case of slight contamination on the floor a wet medium such as dampened sawdust sprinkled over the contaminated area before brushing is acceptable. For all other surfaces, wet methods such as swabbing are essential. The removal of contamination should be done with the minimum of rubbing and the swabs should be frequently discarded as radioactive waste. Decontamination solutions which contain complexing agents are particularly useful in such cases.

Where there is copious loose contamination, a suitable strippable lacquer may be carefully applied to the contaminated surfaces. This lacquer is allowed to dry and in so doing will take up the contamination. Afterwards, the strippable lacquer can be removed together with the contamination. During this operation, care should be taken in using the spraying device to avoid disturbing the loose contamination and thus giving rise to an airborne hazard. As a precaution, personnel should be in fully protective clothing. After stripping, the affected areas should be washed.

ACCIDENTS AND EMERGENCY PROCEDURES

A radiation accident results from the loss of control over a radiation source which could directly or indirectly involve hazards to life, health and property. Radiation accident would normally conform to one of the following broad general patterns:

- Accidental external exposure to excessive amounts of radiation, as for example, when a person inadvertently remains close to a strong source or accidentally gets exposed to a beam of radiation.
- Accidental spill or explosion in a working place (accompanied by a fire or otherwise) resulting in surface and air contamination of the surroundings and contamination of personnel. In such cases, the intake of radioactive substances into the body could be by inhalation, through open wounds resulting from the accident, or by direct absorption through the skin.
- Dispersal of radioactive material to the environment as a result of an explosion, fire, mechanical shock or other incident occurring in a public place, as for instance, when radioactive substances are being transported.

High levels of exposure to radiation occur as a result of a person inadvertently entering a high radiation field such as, for example, walking through an X-ray beam when the machine is on, or when a person gets dangerously close to an industrial source while it is being used for panoramic exposures. Such accidental exposures can also occur when shielding measures are inadequate. In places where a potential for such accidents exists, appropriate control measures should be taken well in advance to ensure that the chances of any person being accidentally exposed to high levels of radiation are minimized. Such measures include the provision of radiation-actuated and other similar interlocks which would ensure that no person can enter a radiation area when an exposure is in progress. Also, provisions of visual or aural indication to identify high radiation level areas and provision of adequate radiation alarms which can be either located at strategic points or carried by individuals whenever they are near high radiation level areas must be included.

A second category of accidental radiation exposure, which could arise in a radiation work area, could result from one of the following contingencies:

- The inability to get a remotely controlled source back into its shielded container because of mechanical or pneumatic failure.
- Accidental breakage of a sealed source or the container of an open source, resulting in high contamination of both surface and air in the vicinity.

- Breakdown of crucial ventilation systems in areas where open sources are being handled.

- Accidents which could involve fire or explosion and which could result in the breakdown of the integrity of shielding, or dispersal of radionuclides in the environment of the laboratory, e.g., in the case of a major nuclear facility, a criticality accident.

Off-site accidents could occur in areas to which the public have access and may result from one of the following contingencies:

- Unplanned release of airborne radioactivity to the environment from a radiation facility due to unusual conditions such as fire, explosion, breakdown of the ventilation system or breakdown of the filter system.

- Accident to consignments of radionuclides when such consignments are in a carrier such as a truck, train or aircraft, or when such consignments are held in storage during transit.

The first essential step in dealing with a radiation accident is to identify, segregate and treat all persons who may have been subjected to radiation exposures both external and internal. Immediate steps should be taken to assess the extent of exposure by sending the personnel monitoring devices used by the exposed persons for immediate processing. Biological monitoring and body burden measurements must also be immediately conducted where necessary.

The formulation of any radiation safety program must first be aimed at the determination of the nature and extent of the health hazards involved in any operation with radiation sources. On the basis of such an assessment, the installation has to be appropriately planned and operating procedures have to be formulated so that the levels both for radiation dose and for concentrations of radionuclides in air and water will be kept as low as practicable and in any case, will not exceed the maximum permissible values. The International Commission on Radiological Protection (ICRP) has been active in establishing and keeping under review such maximum permissible values. These values have been identified both for occupational workers and for members of the public, and have been arrived at on the basis that the risk of damage to the health of the persons exposed to radiation at levels not greater than these maximum permissible values is considered to be negligible and, hence, acceptable. The values are widely accepted as determining the basis for the design of installations, the development of suitable operational procedures and the formulation of appropriate administrative constraints. However, the ICRP has also recommended that all radiation exposures should be kept at the lowest practicable limits. Thus, while maximum permissible levels can serve as a guide for planning purposes, they should not be treated as levels to which radiation workers could be routinely exposed.

The organization in charge of an overall radiation safety program should be responsible for:

1. The formulation and implementation of appropriate radiation protection regulations.
2. The siting, location and design of radiation installations with particular reference to: (1) the types of radiation sources to be used, (2) environmental factors related to the disposal of radioactive material and to its dispersal both under normal and emergency conditions, and (3) the presence of occupied areas in the vicinity of the installation.
3. Structural design features which would have a bearing on (1) the possible spread of contamination throughout the area, and (2) ease of decontamination.
4. The setting up of house rules and well-defined operational procedures.
5. The proper instruction of personnel in these rules and procedures.
6. The provision of all necessary facilities for (1) personnel monitoring, (2) area monitoring, (3) medical supervison and (4) the maintenance of all the relevant records.
7. The drawing up of procedures for meeting emergencies and the provision of all the facilities necessary for carrying out these procedures.
8. The maintenance of proper liaison with external agencies, such as the fire, police, transport, and public health authorities.
9. The maintenance of cumulative whole-body radiation exposure records covering both internal and external exposures.
10. The initiation of appropriate action in cases of excessive exposures or radiation emergencies.

CONSIDERATIONS FOR TRANSPORT

Radioactive materials are being regularly transported by land, water or air in compliance with the current national or international regulations. Even during normal conditions of transport and handling, there exist possibilities of radiation exposure of (a) personnel involved in such transport (crew, passengers, etc.), and (b) undeveloped photographic films that may be contiguously present in the means of transport. Further, the release of radioactive materials to the environment under accidental conditions cannot be ignored. To minimize these hazards and to ensure safe transport of radioactive materials several steps are taken, which include (1) limiting the amount of radioactive material in a package, depending on the ability of the package to withstand both normal and abnormal conditions encountered during transport, (2) limiting the radiation level on the surface of the package and at a distance of 1 meter from the surface of the package, and (3) segregating such packages from passenger areas and undeveloped photographic films. Based on these considerations, national and international regulations governing the transport of radioactive materials have been formulated.

The term package refers to the packaging together with its radioactive contents as presented for transport. Packaging is the assembly of components necessary to ensure compliance with the packaging requirements of the regulations. It may, in particular, consist of one or more receptacles, absorbent materials, spacing structures, radiation shielding, and devices for cooling, for absorbing mechanical shocks and for thermal insulation. These devices may include the vehicle with tie-down system when these are intended to form an integral part of the packaging. The two types of packaging are Type A and Type B. Type A packaging is designed to withstand the normal conditions of transport and is required to demonstrate its suitability by the retention of integrity of shielding and containment under normal conditions incident to transport. Type B packaging is designed to withstand the damaging effects of a transport accident as demonstrated by the retention of integrity of shielding and containment. The limits of activity permitted for Type B packagings are higher than for Type A packagings. Type A package is a Type A packaging together with its limited radioactive contents and does not require competent authority approval. Type B(U) package is a Type B packaging together with its radioactive contents, which requires unilateral approval only of the package design and of any stowage provisions that may be necessary for heat dissipation. Type B(M) package is a Type B packaging together with its radioactive contents as presented for transport and which requires multilateral approval of the package design and of the conditions of shipment.

Packages are further classified as Category I-White, Category II-Yellow, and Category III-Yellow packages based on the radiation levels on their surface and at one meter from any point on the surface. While most of the pure alpha and beta emitters could be transported as White packages, it would be economical to transport gamma emitters as Yellow packages. Otherwise, considerable amounts of shielding would be required to bring the radiation levels to those corresponding to White packages. Each package has to be appropriately labelled and the activity limits as well as labelling requirements have to follow the relevent transport regulations. Packages containing radioactive materials which are also fissile materials are designed, except for certain specified cases, to comply with the general provisions for nuclear safety. All fissile materials are packed and shipped in such a manner that criticality cannot be reached under any foreseeable circumstances of transport.

Packages of fissile materials are divided into: Fissile Class I: packages which are nuclearly safe in any number and in any arrangement under all foreseeable circumstances of transport; Fissile Class II: packages which in limited number are nuclearly safe in any arrangement under all foreseeable circumstances of transport; Fissile Class III: packages which are nuclearly safe under all foreseeable circumstances of transport by reason of special precautions, or special administrative or operational controls imposed upon the transport of the consignment.

REFERENCES

1 Alexander, P. *Atomic Radiation and Life*. Revised edition. Baltimore:Penguin Books, Inc. (1965).

2 Attix, F. H, and W. C. Roesch, eds. *Radiation Dosimetry, Vol. 1*. 2nd edition. New York:Academic Press (1968).

3 Attix, F. H. and E. Tochilin, eds. *Radiation Dosimetry, Vol. 3*. 2nd edition. New York:Academic Press (1969).

4 Glastone, S. *Source Book on Atomic Energy*. London:D. Van Nostrand Co., Inc. (1967).

5 International Atomic Energy Agency. "Disposal of Radioactive Wastes into Rivers, Lakes and Estuaries," Safety Series No. 36, IAEA, Vienna (1971).

6 International Atomic Energy Agency. "Radiation Protection Procedures," IAEA, Austria (1973).

APPENDIX

MAXIMUM PERMISSIBLE AVERAGE CONCENTRATIONS OF RADIOACTIVE MATERIALS IN AIR AND WATER

The following is a listing of the Maximum Permissible Average Concentrations of radioactive materials in air and water, taken from the *New Jersey Radiation Protection Code* (RH-D20, Jan. 1969).

Radionuclide		Occupational 40-hr. Week		Non-Occupational	
		Water uc/ml	Air uc/ml	Water uc/ml	Air uc/ml
Column		A	B	C	D
Actinium 227	(sol.)	6×10^{-5}	2×10^{-12}	2×10^{-6}	8×10^{-14}
	(insol.)	9×10^{-3}	3×10^{-11}	3×10^{-4}	9×10^{-18}
Actinium 228	(sol.)	3×10^{-3}	8×10^{-8}	9×10^{-5}	3×10^{-9}
	(insol.)	3×10^{-3}	2×10^{-8}	9×10^{-5}	6×10^{-10}
Americium 241	(sol.)	10^{-4}	6×10^{-12}	4×10^{-6}	2×10^{-13}
	(insol.)	8×10^{-4}	10^{-10}	2×10^{-5}	4×10^{-12}
Americium 242m	(sol.)	1×10^{-4}	6×10^{-12}	4×10^{-6}	2×10^{-13}
	(insol.)	3×10^{-3}	3×10^{-10}	9×10^{-5}	9×10^{-12}
Americium 242	(sol.)	4×10^{-3}	4×10^{-8}	1×10^{-4}	1×10^{-9}
	(insol.)	4×10^{-3}	5×10^{-8}	1×10^{-4}	2×10^{-9}
Americium 243	(sol.)	10^{-4}	6×10^{-12}	4×10^{-6}	2×10^{-13}
	(insol.)	8×10^{-4}	10^{-10}	3×10^{-5}	4×10^{-12}

(continued)

Radionuclide		Occupational 40-hr. Week		Non-Occupational	
		Water uc/ml	Air uc/ml	Water uc/ml	Air uc/ml
Column		A	B	C	D
Americium 244	(sol.)	1×10^{-1}	4×10^{-6}	5×10^{-3}	1×10^{-7}
	(insol.)	1×10^{-1}	2×10^{-5}	5×10^{-3}	8×10^{-7}
Antimony 122	(sol.)	8×10^{-4}	2×10^{-7}	3×10^{-5}	6×10^{-9}
	(insol.)	8×10^{-4}	10^{-7}	3×10^{-5}	5×10^{-9}
Antimony 124	(sol.)	7×10^{-4}	2×10^{-7}	2×10^{-5}	5×10^{-9}
	(insol.)	7×10^{-4}	2×10^{-8}	2×10^{-5}	7×10^{-10}
Antimony 125	(sol.)	3×10^{-3}	5×10^{-7}	10^{-4}	2×10^{-8}
	(insol.)	3×10^{-3}	3×10^{-8}	10^{-4}	9×10^{-10}
Argon 37	(imm.)	....	6×10^{-3}	....	10^{-4}
Argon 41	(imm.)	....	2×10^{-6}	....	4×10^{-8}
Arsenic 73	(sol.)	0.01	2×10^{-6}	5×10^{-4}	7×10^{-8}
	(insol.)	0.01	4×10^{-7}	5×10^{-4}	10^{-8}
Arsenic 74	(sol.)	2×10^{-3}	3×10^{-7}	5×10^{-5}	10^{-8}
	(insol.)	2×10^{-3}	10^{-7}	5×10^{-5}	4×10^{-9}
Arsenic 76	(sol.)	6×10^{-4}	10^{-7}	2×10^{-5}	4×10^{-9}
	(insol.)	6×10^{-4}	10^{-7}	2×10^{-5}	3×10^{-9}
Arsenic 77	(sol.)	2×10^{-3}	5×10^{-7}	8×10^{-5}	2×10^{-8}
	(insol.)	2×10^{-3}	4×10^{-7}	8×10^{-5}	10^{-8}
Astatine 211	(sol.)	5×10^{-5}	7×10^{-9}	2×10^{-6}	2×10^{-10}
	(insol.)	2×10^{-3}	3×10^{-8}	7×10^{-5}	10^{-9}
Barium 131	(sol.)	5×10^{-3}	10^{-6}	2×10^{-4}	4×10^{-8}
	(insol.)	5×10^{-3}	4×10^{-7}	2×10^{-4}	10^{-8}
Barium 140	(sol.)	8×10^{-4}	10^{-7}	3×10^{-5}	4×10^{-9}
	(insol.)	7×10^{-4}	4×10^{-8}	2×10^{-5}	10^{-9}
Berkelium 249	(sol.)	0.02	9×10^{-10}	6×10^{-4}	3×10^{-11}
	(insol.)	0.02	10^{-7}	6×10^{-4}	4×10^{-9}
Berkelium 250	(sol.)	6×10^{-3}	1×10^{-7}	2×10^{-4}	5×10^{-9}
	(insol.)	6×10^{-3}	1×10^{-6}	2×10^{-4}	4×10^{-8}
Beryllium 7	(sol.)	0.05	6×10^{-6}	0.002	2×10^{-7}
	(insol.)	0.05	10^{-6}	0.002	4×10^{-8}
Bismuth 206	(sol.)	10^{-3}	2×10^{-7}	4×10^{-5}	6×10^{-9}
	(insol.)	10^{-3}	10^{-7}	4×10^{-5}	5×10^{-9}
Bismuth 207	(sol.)	2×10^{-3}	2×10^{-7}	6×10^{-5}	6×10^{-9}
	(insol.)	2×10^{-3}	10^{-8}	6×10^{-5}	5×10^{-10}
Bismuth 210	(sol.)	10^{-3}	6×10^{-9}	4×10^{-5}	2×10^{-10}
	(insol.)	10^{-3}	6×10^{-9}	4×10^{-5}	2×10^{-10}
Bismuth 212	(sol.)	0.01	10^{-7}	4×10^{-4}	3×10^{-9}
	(insol.)	0.01	2×10^{-7}	4×10^{-4}	7×10^{-9}

(continued)

Radionuclide		Occupational 40-hr. Week		Non-Occupational	
		Water uc/ml	Air uc/ml	Water uc/ml	Air uc/ml
Column		A	B	C	D
Bromine 82	(sol.)	8×10^{-3}	10^{-6}	3×10^{-4}	4×10^{-8}
	(insol.)	10^{-3}	2×10^{-7}	4×10^{-5}	6×10^{-9}
Cadmium 109	(sol.)	5×10^{-3}	5×10^{-8}	2×10^{-4}	2×10^{-9}
	(insol.)	5×10^{-3}	7×10^{-8}	2×10^{-4}	3×10^{-9}
Cadmium 115m	(sol.)	7×10^{-4}	4×10^{-8}	3×10^{-5}	10^{-9}
	(insol.)	7×10^{-4}	4×10^{-8}	3×10^{-5}	10^{-9}
Cadmium 115	(sol.)	10^{-3}	2×10^{-7}	3×10^{-5}	8×10^{-9}
	(insol.)	10^{-3}	2×10^{-7}	4×10^{-5}	6×10^{-9}
Calcium 45	(sol.)	3×10^{-4}	3×10^{-8}	9×10^{-6}	10^{-9}
	(insol.)	5×10^{-3}	10^{-7}	2×10^{-4}	4×10^{-9}
Calcium 47	(sol.)	10^{-3}	2×10^{-7}	5×10^{-5}	6×10^{-9}
	(insol.)	10^{-3}	2×10^{-7}	3×10^{-5}	6×10^{-9}
Californium 249	(sol.)	10^{-4}	2×10^{-12}	4×10^{-6}	5×10^{-14}
	(insol.)	7×10^{-4}	10^{-10}	2×10^{-5}	3×10^{-12}
Californium 250	(sol.)	4×10^{-4}	5×10^{-12}	10^{-5}	2×10^{-13}
	(insol.)	7×10^{-4}	10^{-10}	3×10^{-5}	3×10^{-12}
Californium 251	(sol.)	1×10^{-4}	2×10^{-12}	4×10^{-6}	6×10^{-14}
	(insol.)	8×10^{-4}	1×10^{-10}	3×10^{-5}	3×10^{-12}
Californium 252	(sol.)	7×10^{-4}	2×10^{-11}	2×10^{-5}	7×10^{-13}
	(insol.)	7×10^{-4}	10^{-10}	2×10^{-5}	4×10^{-12}
Californium 253	(sol.)	4×10^{-3}	8×10^{-10}	1×10^{-4}	3×10^{-11}
	(insol.)	4×10^{-3}	8×10^{-10}	1×10^{-4}	3×10^{-11}
Californium 254	(sol).	4×10^{-6}	5×10^{-12}	10^{-7}	2×10^{-13}
	(insol.)	4×10^{-6}	5×10^{-12}	10^{-7}	2×10^{-13}
Carbon 14	(sol.)	0.02	4×10^{-6}	8×10^{-4}	10^{-7}
	(insol.)	....	5×10^{-5}	....	10^{-6}
Cerium 141	(sol.)	3×10^{-3}	4×10^{-7}	9×10^{-5}	2×10^{-8}
	(insol.)	3×10^{-3}	2×10^{-7}	9×10^{-5}	5×10^{-9}
Cerium 143	(sol.)	10^{-3}	3×10^{-7}	4×10^{-5}	9×10^{-9}
	(insol.)	10^{-3}	2×10^{-7}	4×10^{-5}	7×10^{-9}
Cerium 144	(sol.)	3×10^{-4}	10^{-8}	10^{-5}	3×10^{-10}
	(insol.)	3×10^{-4}	6×10^{-9}	10^{-5}	2×10^{-10}
Cesium 131	(sol.)	0.07	10^{-5}	0.002	4×10^{-7}
	(insol.)	0.03	3×10^{-6}	9×10^{-4}	10^{-7}
Cesium 134m	(sol.)	0.2	4×10^{-5}	0.006	10^{-6}
	(insol.)	0.03	6×10^{-6}	0.001	2×10^{-7}
Cesium 134	(sol.)	3×10^{-4}	4×10^{-8}	9×10^{-6}	10^{-9}
	(insol.)	10^{-3}	10^{-8}	4×10^{-5}	4×10^{-10}

(continued)

Radionuclide		Occupational 40-hr. Week		Non-Occupational	
		Water uc/ml	Air uc/ml	Water uc/ml	Air uc/ml
Column		A	B	C	D
Cesium 135	(sol.)	3×10^{-3}	5×10^{-7}	10^{-4}	2×10^{-8}
	(insol.)	7×10^{-3}	9×10^{-8}	2×10^{-4}	3×10^{-9}
Cesium 136	(sol.)	2×10^{-3}	4×10^{-7}	9×10^{-5}	10^{-8}
	(insol.)	2×10^{-3}	2×10^{-7}	6×10^{-5}	6×10^{-9}
Cesium 137	(sol.)	4×10^{-4}	6×10^{-8}	2×10^{-5}	2×10^{-9}
	(insol.)	10^{-3}	10^{-8}	4×10^{-5}	5×10^{-10}
Chlorine 36	(sol.)	2×10^{-3}	4×10^{-7}	8×10^{-5}	10^{-8}
	(insol.)	2×10^{-3}	2×10^{-8}	6×10^{-5}	8×10^{-10}
Chlorine 38	(sol.)	0.01	3×10^{-6}	4×10^{-4}	9×10^{-8}
	(insol.)	0.01	2×10^{-6}	4×10^{-4}	7×10^{-8}
Chromium 51	(sol.)	0.05	10^{-5}	0.002	4×10^{-7}
	(insol.)	0.05	2×10^{-6}	0.002	8×10^{-8}
Cobalt 57	(sol.)	0.02	3×10^{-6}	5×10^{-4}	10^{-7}
	(insol.)	0.01	2×10^{-7}	4×10^{-4}	6×10^{-9}
Cobalt 58m	(sol.)	0.08	2×10^{-5}	0.003	6×10^{-7}
	(insol.)	0.06	9×10^{-6}	0.002	3×10^{-7}
Cobalt 58	(sol.)	4×10^{-3}	8×10^{-7}	10^{-4}	3×10^{-8}
	(insol.)	3×10^{-3}	5×10^{-8}	9×10^{-5}	2×10^{-9}
Cobalt 60	(sol.)	10^{-3}	3×10^{-7}	5×10^{-5}	10^{-8}
	(insol.)	10^{-3}	9×10^{-9}	3×10^{-5}	3×10^{-10}
Copper 64	(sol.)	0.01	2×10^{-6}	3×10^{-4}	7×10^{-8}
	(insol.)	6×10^{-3}	10^{-6}	2×10^{-4}	4×10^{-8}
Curium 242	(sol.)	7×10^{-4}	10^{-10}	2×10^{-5}	4×10^{-12}
	(insol.)	7×10^{-4}	2×10^{-10}	3×10^{-5}	6×10^{-12}
Curium 243	(sol.)	10^{-4}	6×10^{-12}	5×10^{-6}	2×10^{-13}
	(insol.)	7×10^{-4}	10^{-10}	2×10^{-5}	3×10^{-12}
Curium 244	(sol.)	2×10^{-4}	9×10^{-12}	7×10^{-6}	3×10^{-13}
	(insol.)	8×10^{-4}	10^{-10}	3×10^{-5}	3×10^{-12}
Curium 245	(sol.)	10^{-4}	5×10^{-12}	4×10^{-6}	2×10^{-13}
	(insol.)	8×10^{-4}	10^{-10}	3×10^{-5}	4×10^{-12}
Curium 246	(sol.)	10^{-4}	5×10^{-12}	4×10^{-6}	2×10^{-13}
	(insol.)	8×10^{-4}	10^{-10}	3×10^{-5}	4×10^{-12}
Curium 247	(sol.)	1×10^{-4}	5×10^{-12}	4×10^{-6}	2×10^{-13}
	(insol.)	6×10^{-4}	1×10^{-10}	2×10^{-5}	4×10^{-12}
Curium 248	(sol.)	1×10^{-5}	6×10^{-13}	4×10^{-7}	2×10^{-14}
	(insol.)	4×10^{-5}	1×10^{-11}	1×10^{-6}	4×10^{-13}
Curium 249	(sol.)	6×10^{-2}	1×10^{-5}	2×10^{-3}	4×10^{-7}
	(insol.)	6×10^{-2}	1×10^{-5}	2×10^{-3}	4×10^{-7}

(continued)

Radionuclide		Occupational 40-hr. Week		Non-Occupational	
		Water uc/ml	Air uc/ml	Water uc/ml	Air uc/ml
Column		A	B	C	D
Dysprosium 165	(sol.)	0.01	3×10^{-6}	4×10^{-4}	9×10^{-8}
	(insol.)	0.01	2×10^{-6}	4×10^{-4}	7×10^{-8}
Dysprosium 166	(sol.)	10^{-3}	2×10^{-7}	4×10^{-5}	8×10^{-9}
	(insol.)	10^{-3}	2×10^{-7}	4×10^{-5}	7×10^{-9}
Einsteinium 253	(sol.)	7×10^{-4}	8×10^{-10}	2×10^{-5}	3×10^{-11}
	(insol.)	7×10^{-4}	6×10^{-10}	2×10^{-5}	2×10^{-11}
Einsteinium 254m	(sol.)	5×10^{-4}	5×10^{-9}	2×10^{-5}	2×10^{-10}
	(insol.)	5×10^{-4}	6×10^{-9}	2×10^{-5}	2×10^{-10}
Einsteinium 254	(sol.)	4×10^{-4}	2×10^{-11}	1×10^{-5}	6×10^{-13}
	(insol.)	4×10^{-4}	1×10^{-10}	1×10^{-5}	4×10^{-12}
Einsteinium 255	(sol.)	8×10^{-4}	5×10^{-10}	3×10^{-5}	2×10^{-11}
	(insol.)	8×10^{-4}	4×10^{-10}	3×10^{-5}	1×10^{-11}
Erbium 169	(sol.)	3×10^{-3}	6×10^{-7}	9×10^{-5}	2×10^{-8}
	(insol.)	3×10^{-3}	4×10^{-7}	9×10^{-5}	10^{-8}
Erbium 171	(sol.)	3×10^{-3}	7×10^{-7}	10^{-4}	2×10^{-8}
	(insol.)	3×10^{-3}	6×10^{-7}	10^{-4}	2×10^{-8}
Europium 152	(9.2 hr).				
	(sol.)	2×10^{-3}	4×10^{-7}	6×10^{-5}	10^{-8}
	(insol.)	2×10^{-3}	3×10^{-1}	6×10^{-5}	10^{-8}
Europium 152	(13 yr.)				
	(sol.)	2×10^{-3}	10^{-8}	8×10^{-5}	4×10^{-10}
	(insol.)	2×10^{-3}	2×10^{-8}	8×10^{-5}	6×10^{-10}
Europium 154	(sol.)	6×10^{-4}	4×10^{-9}	2×10^{-5}	10^{-10}
	(insol.)	6×10^{-4}	7×10^{-9}	2×10^{-5}	2×10^{-10}
Europium 155	(sol.)	6×10^{-3}	9×10^{-8}	2×10^{-4}	3×10^{-9}
	(insol.)	6×10^{-3}	7×10^{-8}	2×10^{-4}	3×10^{-9}
Fermium 254	(sol.)	4×10^{-3}	6×10^{-8}	1×10^{-4}	2×10^{-9}
	(insol.)	4×10^{-3}	7×10^{-8}	1×10^{-4}	2×10^{-9}
Fermium 255	(sol.)	1×10^{-3}	2×10^{-8}	3×10^{-5}	6×10^{-10}
	(insol.)	1×10^{-3}	1×10^{-8}	3×10^{-5}	4×10^{-10}
Fermium 256	(sol.)	3×10^{-5}	3×10^{-9}	9×10^{-7}	1×10^{-10}
	(insol.)	3×10^{-5}	2×10^{-9}	9×10^{-7}	6×10^{-11}
Fluorine 18	(sol.)	0.02	5×10^{-6}	8×10^{-4}	2×10^{-7}
	(insol.)	0.01	3×10^{-6}	5×10^{-4}	9×10^{-8}
Gadolinium 153	(sol.)	6×10^{-3}	2×10^{-7}	2×10^{-4}	8×10^{-9}
	(insol.)	6×10^{-3}	9×10^{-8}	2×10^{-4}	3×10^{-9}
Gadolinium 159	(sol.)	2×10^{-3}	5×10^{-7}	8×10^{-5}	2×10^{-8}
	(insol.)	2×10^{-3}	4×10^{-7}	8×10^{-5}	10^{-8}

(continued)

Radionuclide		Occupational 40-hr. Week		Non-Occupational	
		Water uc/ml	Air uc/ml	Water uc/ml	Air uc/ml
Column		A	B	C	D
Gallium 72	(sol.)	10^{-3}	2×10^{-7}	4×10^{-5}	8×10^{-9}
	(insol.)	10^{-3}	2×10^{-7}	4×10^{-5}	6×10^{-9}
Germanium 71	(sol.)	0.05	10^{-5}	0.002	4×10^{-7}
	(insol.)	0.05	6×10^{-6}	0.002	2×10^{-7}
Gold 196	(sol.)	5×10^{-3}	10^{-6}	2×10^{-4}	4×10^{-8}
	(insol.)	4×10^{-3}	6×10^{-7}	10^{-4}	2×10^{-8}
Gold 198	(sol.)	2×10^{-3}	3×10^{-7}	5×10^{-5}	10^{-8}
	(insol.)	10^{-3}	2×10^{-7}	5×10^{-5}	8×10^{-9}
Gold 199	(sol.)	5×10^{-3}	10^{-6}	2×10^{-4}	4×10^{-8}
	(insol.)	4×10^{-3}	8×10^{-7}	2×10^{-4}	3×10^{-8}
Hafnium 181	(sol.)	2×10^{-3}	4×10^{-8}	7×10^{-5}	10^{-9}
	(insol.)	2×10^{-3}	7×10^{-8}	7×10^{-5}	3×10^{-9}
Holmium 166	(sol.)	9×10^{-4}	2×10^{-7}	3×10^{-5}	7×10^{-9}
	(insol.)	9×10^{-4}	2×10^{-7}	3×10^{-5}	6×10^{-9}
Hydrogen 3	(sol., insol.)	0.1	5×10^{-6}	0.003	2×10^{-7}
	(imm.)	....	2×10^{-3}	....	4×10^{-5}
Indium 113m	(sol.)	0.04	8×10^{-6}	0.001	3×10^{-7}
	(insol.)	0.04	7×10^{-6}	0.001	2×10^{-7}
Indium 114m	(sol.)	5×10^{-4}	10^{-7}	2×10^{-5}	4×10^{-9}
	(insol.)	5×10^{-4}	2×10^{-8}	2×10^{-5}	7×10^{-10}
Indium 115m	(sol.)	0.01	2×10^{-6}	4×10^{-4}	8×10^{-8}
	(insol.)	0.01	2×10^{-6}	4×10^{-4}	6×10^{-8}
Indium 115	(sol.)	3×10^{-3}	2×10^{-7}	9×10^{-5}	9×10^{-9}
	(insol.)	3×10^{-3}	3×10^{-8}	9×10^{-5}	10^{-9}
Iodine 125	(sol.)	4×10^{-5}	5×10^{-9}	2×10^{-7}	8×10^{-11}
	(insol.)	6×10^{-3}	2×10^{-7}	2×10^{-4}	6×10^{-9}
Iodine 126	(sol.)	5×10^{-5}	8×10^{-9}	3×10^{-7}	9×10^{-11}
	(insol.)	3×10^{-3}	3×10^{-7}	9×10^{-5}	10^{-8}
Iodine 129	(sol.)	10^{-5}	2×10^{-9}	6×10^{-8}	2×10^{-11}
	(insol.)	6×10^{-3}	7×10^{-8}	2×10^{-4}	2×10^{-9}
Iodine 131	(sol.)	6×10^{-5}	9×10^{-9}	3×10^{-7}	1×10^{-10}
	(insol.)	2×10^{-3}	3×10^{-7}	6×10^{-5}	10^{-8}
Iodine 132	(sol.)	2×10^{-3}	2×10^{-7}	8×10^{-6}	3×10^{-9}
	(insol.)	5×10^{-3}	9×10^{-7}	2×10^{-4}	3×10^{-8}
Iodine 133	(sol.)	2×10^{-4}	3×10^{-8}	1×10^{-6}	4×10^{-10}
	(insol.)	10^{-3}	2×10^{-7}	4×10^{-5}	7×10^{-9}
Iodine 134	(sol.)	4×10^{-3}	5×10^{-7}	2×10^{-5}	6×10^{-9}
	(insol.)	0.02	3×10^{-6}	6×10^{-4}	10^{-7}

(continued)

Radionuclide		Occupational 40-hr. Week		Non-Occupational	
		Water uc/ml	Air uc/ml	Water uc/ml	Air uc/ml
Column		A	B	C	D
Iodine 135	(sol.)	7×10^{-4}	10^{-7}	4×10^{-6}	1×10^{-9}
	(insol.)	2×10^{-3}	4×10^{-7}	7×10^{-5}	10^{-8}
Iridium 190	(sol.)	6×10^{-3}	10^{-6}	2×10^{-4}	4×10^{-8}
	(insol.)	5×10^{-3}	4×10^{-7}	2×10^{-4}	10^{-8}
Iridium 192	(sol.)	10^{-3}	10^{-7}	4×10^{-5}	4×10^{-9}
	(insol.)	10^{-3}	3×10^{-8}	4×10^{-5}	9×10^{-10}
Iridium 194	(sol.)	10^{-3}	2×10^{-7}	3×10^{-5}	8×10^{-9}
	(insol.)	9×10^{-4}	2×10^{-7}	3×10^{-5}	5×10^{-9}
Iron 55	(sol.)	0.02	9×10^{-7}	8×10^{-4}	3×10^{-8}
	(insol.)	0.07	10^{-6}	0.002	3×10^{-8}
Iron 59	(sol.)	2×10^{-3}	10^{-7}	6×10^{-5}	5×10^{-9}
	(insol.)	2×10^{-3}	5×10^{-8}	5×10^{-5}	2×10^{-9}
Krypton 85m	(imm.)	....	6×10^{-6}	....	10^{-7}
Krypton 85	(imm.)	....	10^{-5}	....	3×10^{-7}
Krypton 87	(imm.)	....	10^{-6}	....	2×10^{-8}
Lanthanum 140	(sol.)	7×10^{-4}	2×10^{-7}	2×10^{-5}	5×10^{-9}
	(insol.)	7×10^{-4}	10^{-7}	2×10^{-5}	4×10^{-9}
Lead 203	(sol.)	0.01	3×10^{-6}	4×10^{-4}	9×10^{-8}
	(insol.)	0.01	2×10^{-6}	4×10^{-4}	6×10^{-8}
Lead 210	(sol.)	4×10^{-6}	10^{-10}	10^{-7}	4×10^{-12}
	(insol.)	5×10^{-3}	2×10^{-10}	2×10^{-4}	8×10^{-12}
Lead 212	(sol.)	6×10^{-4}	2×10^{-8}	2×10^{-5}	6×10^{-10}
	(insol.)	5×10^{-4}	2×10^{-8}	2×10^{-5}	7×10^{-10}
Lutecium 177	(sol.)	3×10^{-3}	6×10^{-7}	10^{-4}	2×10^{-8}
	(insol.)	3×10^{-3}	5×10^{-7}	10^{-4}	2×10^{-8}
Manganese 52	(sol.)	10^{-3}	2×10^{-7}	3×10^{-5}	7×10^{-9}
	(insol.)	9×10^{-4}	10^{-7}	3×10^{-5}	5×10^{-9}
Manganese 54	(sol.)	4×10^{-3}	4×10^{-7}	10^{-4}	10^{-8}
	(insol.)	3×10^{-3}	4×10^{-8}	10^{-4}	10^{-9}
Manganese 56	(sol.)	4×10^{-3}	8×10^{-7}	10^{-4}	3×10^{-8}
	(insol.)	3×10^{-3}	5×10^{-7}	10^{-4}	2×10^{-8}
Mercury 197m	(sol.)	6×10^{-3}	7×10^{-7}	2×10^{-4}	3×10^{-8}
	(insol.)	5×10^{-3}	8×10^{-7}	2×10^{-4}	3×10^{-8}
Mercury 197	(sol.)	9×10^{-3}	10^{-6}	3×10^{-4}	4×10^{-8}
	(insol.)	0.01	3×10^{-6}	5×10^{-4}	9×10^{-8}
Mercury 203	(sol.)	5×10^{-4}	7×10^{-8}	2×10^{-5}	2×10^{-9}
	(insol.)	3×10^{-3}	10^{-7}	10^{-4}	4×10^{-9}
Molybdenum 99	(sol.)	5×10^{-3}	7×10^{-7}	2×10^{-4}	3×10^{-8}
	(insol.)	10^{-3}	2×10^{-7}	4×10^{-5}	7×10^{-9}

(continued)

Radionuclide		Occupational 40-hr. Week		Non-Occupational	
		Water uc/ml	Air uc/ml	Water uc/ml	Air uc/ml
Column		A	B	C	D
Neodymium 144	(sol.)	2×10^{-3}	8×10^{-11}	7×10^{-5}	3×10^{-12}
	(insol.)	2×10^{-3}	3×10^{-10}	8×10^{-5}	10^{-11}
Neodymium 147	(sol.)	2×10^{-3}	4×10^{-7}	6×10^{-5}	10^{-8}
	(insol.)	2×10^{-3}	2×10^{-7}	6×10^{-5}	8×10^{-9}
Neodymium 149	(sol.)	8×10^{-3}	2×10^{-6}	3×10^{-4}	6×10^{-8}
	(insol.)	8×10^{-3}	10^{-6}	3×10^{-4}	5×10^{-8}
Neptunium 237	(sol.)	9×10^{-5}	4×10^{-12}	3×10^{-6}	10^{-13}
	(insol.)	9×10^{-4}	10^{-10}	3×10^{-5}	4×10^{-12}
Neptunium 239	(sol.)	4×10^{-3}	8×10^{-7}	10^{-4}	3×10^{-8}
	(insol.)	4×10^{-3}	7×10^{-7}	10^{-4}	2×10^{-8}
Nickel 59	(sol.)	6×10^{-3}	5×10^{-7}	2×10^{-4}	2×10^{-8}
	(insol.)	0.06	8×10^{-7}	0.002	3×10^{-8}
Nickel 63	(sol.)	8×10^{-4}	6×10^{-8}	3×10^{-5}	2×10^{-9}
	(insol.)	0.02	3×10^{-7}	7×10^{-4}	10^{-8}
Nickel 65	(sol.)	4×10^{-3}	9×10^{-7}	10^{-4}	3×10^{-8}
	(insol.)	3×10^{-3}	5×10^{-7}	10^{-4}	2×10^{-8}
Niobium 93m	(sol.)	0.01	10^{-7}	4×10^{-4}	4×10^{-9}
	(insol.)	0.01	2×10^{-7}	4×10^{-4}	5×10^{-9}
Niobium 95	(sol.)	3×10^{-3}	5×10^{-7}	10^{-4}	2×10^{-8}
	(insol.)	3×10^{-3}	10^{-7}	10^{-4}	3×10^{-9}
Niobium 97	(sol.)	0.03	6×10^{-6}	9×10^{-4}	2×10^{-7}
	(insol.)	0.03	5×10^{-6}	9×10^{-4}	2×10^{-7}
Osmium 185	(sol.)	2×10^{-3}	5×10^{-7}	7×10^{-5}	2×10^{-8}
	(insol.)	2×10^{-3}	5×10^{-8}	7×10^{-5}	2×10^{-9}
Osmium 191m	(sol.)	0.07	2×10^{-5}	0.003	6×10^{-7}
	(insol.)	0.07	9×10^{-6}	0.002	3×10^{-7}
Osmium 191	(sol.)	5×10^{-3}	10^{-6}	2×10^{-4}	4×10^{-8}
	(insol.)	5×10^{-3}	4×10^{-7}	2×10^{-4}	10^{-8}
Osmium 193	(sol.)	2×10^{-3}	4×10^{-7}	6×10^{-5}	10^{-8}
	(insol.)	2×10^{-3}	3×10^{-7}	5×10^{-5}	9×10^{-9}
Palladium 103	(sol.)	0.01	10^{-6}	3×10^{-4}	5×10^{-8}
	(insol.)	8×10^{-3}	7×10^{-7}	3×10^{-4}	3×10^{-8}
Palladium 109	(sol.)	3×10^{-3}	6×10^{-7}	9×10^{-5}	2×10^{-8}
	(insol.)	2×10^{-3}	4×10^{-7}	7×10^{-5}	10^{-8}
Phosphorus 32	(sol.)	5×10^{-4}	7×10^{-8}	2×10^{-5}	2×10^{-9}
	(insol.)	7×10^{-4}	8×10^{-8}	2×10^{-5}	3×10^{-9}
Platinum 191	(sol.)	4×10^{-3}	8×10^{-7}	10^{-4}	3×10^{-8}
	(insol.)	3×10^{-3}	6×10^{-7}	10^{-4}	2×10^{-8}

(continued)

Radionuclide		Occupational 40-hr. Week		Non-Occupational	
		Water uc/ml	Air uc/ml	Water uc/ml	Air uc/ml
Column		**A**	**B**	**C**	**D**
Platinum 193m	(sol.)	0.03	7×10^{-6}	0.001	2×10^{-7}
	(insol.)	0.03	5×10^{-6}	0.001	2×10^{-7}
Platinum 193	(sol.)	0.03	10^{-6}	9×10^{-4}	4×10^{-8}
	(insol.)	0.05	3×10^{-7}	0.002	10^{-8}
Platinum 197m	(sol.)	0.03	6×10^{-6}	0.001	2×10^{-7}
	(insol.)	0.03	5×10^{-6}	9×10^{-4}	2×10^{-7}
Platinum 197	(sol.)	4×10^{-3}	8×10^{-7}	10^{-4}	3×10^{-8}
	(insol.)	3×10^{-3}	6×10^{-7}	10^{-4}	2×10^{-8}
Plutonium 238	(sol.)	10^{-4}	2×10^{-12}	5×10^{-6}	7×10^{-14}
	(insol.)	8×10^{-4}	3×10^{-11}	3×10^{-5}	10^{-12}
Plutonium 239	(sol.)	10^{-4}	2×10^{-12}	5×10^{-6}	6×10^{-14}
	(insol.)	8×10^{-4}	4×10^{-11}	3×10^{-5}	10^{-12}
Plutonium 240	(sol.)	10^{-4}	2×10^{-12}	5×10^{-6}	6×10^{-14}
	(insol.)	8×10^{-4}	4×10^{-11}	3×10^{-5}	10^{-12}
Plutonium 241	(sol.)	7×10^{-3}	9×10^{-11}	2×10^{-4}	3×10^{-12}
	(insol.)	0.04	4×10^{-8}	0.001	10^{-9}
Plutonium 242	(sol.)	10^{-4}	2×10^{-12}	5×10^{-6}	6×10^{-14}
	(insol.)	9×10^{-4}	4×10^{-11}	3×10^{-5}	10^{-12}
Plutonium 243	(sol.)	1×10^{-2}	2×10^{-6}	3×10^{-4}	6×10^{-8}
	(insol.)	1×10^{-2}	2×10^{-6}	3×10^{-4}	8×10^{-8}
Plutonium 244	(sol.)	1×10^{-4}	2×10^{-12}	4×10^{-6}	6×10^{-14}
	(insol.)	3×10^{-4}	3×10^{-11}	1×10^{-5}	1×10^{-12}
Polonium 210	(sol.)	2×10^{-5}	5×10^{-10}	7×10^{-7}	2×10^{-11}
	(insol.)	8×10^{-4}	2×10^{-10}	3×10^{-5}	7×10^{-12}
Potassium 42	(sol.)	9×10^{-3}	2×10^{-6}	3×10^{-4}	7×10^{-8}
	(insol.)	6×10^{-4}	10^{-7}	2×10^{-5}	4×10^{-9}
Praseodymium 142	(sol.)	9×10^{-4}	2×10^{-7}	3×10^{-5}	7×10^{-9}
	(insol.)	9×10^{-4}	2×10^{-7}	3×10^{-5}	5×10^{-9}
Praseodymium 143	(sol.)	10^{-3}	3×10^{-7}	5×10^{-5}	10^{-8}
	(insol.)	10^{-3}	2×10^{-7}	5×10^{-5}	6×10^{-9}
Promethium 147	(sol.)	6×10^{-3}	6×10^{-8}	2×10^{-4}	2×10^{-9}
	(insol.)	6×10^{-3}	10^{-7}	2×10^{-4}	3×10^{-9}
Promethium 149	(sol.)	10^{-3}	3×10^{-7}	4×10^{-5}	10^{-8}
	(insol.)	10^{-3}	2×10^{-7}	4×10^{-5}	8×10^{-9}
Protactinium 230	(sol.)	7×10^{-3}	2×10^{-9}	2×10^{-4}	6×10^{-11}
	(insol.)	7×10^{-3}	8×10^{-10}	2×10^{-4}	3×10^{-11}
Protactinium 231	(sol.)	3×10^{-5}	10^{-12}	9×10^{-7}	4×10^{-14}
	(insol.)	8×10^{-4}	10^{-10}	2×10^{-5}	4×10^{-12}

(continued)

Radionuclide		Occupational 40-hr. Week		Non-Occupational	
		Water uc/ml	Air uc/ml	Water uc/ml	Air uc/ml
Column		A	B	C	D
Protactinium 233	(sol.)	4×10^{-3}	6×10^{-7}	10^{-4}	2×10^{-8}
	(insol.)	3×10^{-3}	2×10^{-7}	10^{-4}	6×10^{-9}
Radium 223	(sol.)	2×10^{-5}	2×10^{-9}	7×10^{-7}	6×10^{-11}
	(insol.)	10^{-4}	2×10^{-10}	4×10^{-6}	8×10^{-12}
Radium 224	(sol.)	7×10^{-5}	5×10^{-9}	2×10^{-6}	2×10^{-10}
	(insol.)	2×10^{-4}	7×10^{-10}	5×10^{-6}	2×10^{-11}
Radium 226	(sol.)	4×10^{-7}	3×10^{-11}	3×10^{-8}	3×10^{-12}
	(insol.)	9×10^{-4}	5×10^{-11}	3×10^{-5}	2×10^{-12}
Radium 228	(sol.)	8×10^{-7}	7×10^{-11}	3×10^{-8}	2×10^{-12}
	(insol.)	7×10^{-4}	4×10^{-11}	3×10^{-5}	10^{-12}
Radon 220		....	3×10^{-7}	....	10^{-8}
Radon 222		....	3×10^{-8}	....	1×10^{-9}
Rhenium 183	(sol.)	0.02	3×10^{-6}	6×10^{-4}	9×10^{-8}
	(insol.)	8×10^{-3}	2×10^{-7}	3×10^{-4}	5×10^{-9}
Rhenium 186	(sol.)	3×10^{-3}	6×10^{-7}	9×10^{-5}	2×10^{-8}
	(insol.)	10^{-3}	2×10^{-7}	5×10^{-5}	8×10^{-9}
Rhenium 187	(sol.)	0.07	9×10^{-6}	0.003	3×10^{-7}
	(insol.)	0.04	5×10^{-7}	0.002	2×10^{-8}
Rhenium 188	(sol.)	2×10^{-3}	4×10^{-7}	6×10^{-5}	10^{-8}
	(insol.)	9×10^{-4}	2×10^{-7}	3×10^{-5}	6×10^{-9}
Rhodium 103m	(sol.)	0.4	8×10^{-5}	0.01	3×10^{-6}
	(insol.)	0.3	6×10^{-5}	0.01	2×10^{-6}
Rhodium 105	(sol.)	4×10^{-3}	8×10^{-7}	10^{-4}	3×10^{-8}
	(insol.)	3×10^{-3}	5×10^{-7}	10^{-4}	2×10^{-8}
Rubidium 86	(sol.)	2×10^{-3}	3×10^{-7}	7×10^{-5}	10^{-8}
	(insol.)	7×10^{-4}	7×10^{-8}	2×10^{-5}	2×10^{-9}
Rubidium 87	(sol.)	3×10^{-3}	5×10^{-7}	10^{-4}	2×10^{-8}
	(insol.)	5×10^{-3}	7×10^{-8}	2×10^{-4}	2×10^{-9}
Ruthenium 97	(sol.)	0.01	2×10^{-6}	4×10^{-4}	8×10^{-8}
	(insol.)	0.01	2×10^{-6}	3×10^{-4}	6×10^{-8}
Ruthenium 103	(sol.)	2×10^{-3}	5×10^{-7}	8×10^{-5}	2×10^{-8}
	(insol.)	2×10^{-3}	8×10^{-8}	8×10^{-5}	3×10^{-9}
Ruthenium 105	(sol.)	3×10^{-3}	7×10^{-7}	10^{-4}	2×10^{-8}
	(insol.)	3×10^{-3}	5×10^{-7}	10^{-4}	2×10^{-8}
Ruthenium 106	(sol.)	4×10^{-4}	8×10^{-8}	10^{-5}	3×10^{-9}
	(insol.)	3×10^{-4}	6×10^{-9}	10^{-5}	2×10^{-10}
Samarium 147	(sol.)	2×10^{-3}	7×10^{-11}	6×10^{-5}	2×10^{-12}
	(insol.)	2×10^{-3}	3×10^{-10}	7×10^{-5}	9×10^{-12}

(continued)

Radionuclide		Occupational 40-hr. Week		Non-Occupational	
		Water uc/ml	Air uc/ml	Water uc/ml	Air uc/ml
Column		A	B	C	D
Samarium 151	(sol.)	0.01	6×10^{-8}	4×10^{-4}	2×10^{-9}
	(insol.)	0.01	10^{-7}	4×10^{-4}	5×10^{-9}
Samarium 153	(sol.)	2×10^{-3}	5×10^{-7}	8×10^{-5}	2×10^{-8}
	(insol.)	2×10^{-3}	4×10^{-7}	8×10^{-5}	10^{-8}
Scandium 46	(sol.)	10^{-3}	2×10^{-7}	4×10^{-5}	8×10^{-9}
	(insol.)	10^{-3}	2×10^{-8}	4×10^{-5}	8×10^{-10}
Scandium 47	(sol.)	3×10^{-3}	6×10^{-7}	9×10^{-5}	2×10^{-8}
	(insol.)	3×10^{-3}	5×10^{-7}	9×10^{-5}	2×10^{-8}
Scandium 48	(sol.)	8×10^{-4}	2×10^{-7}	3×10^{-5}	6×10^{-9}
	(insol.)	8×10^{-4}	10^{-7}	3×10^{-5}	5×10^{-9}
Selenium 75	(sol.)	9×10^{-3}	10^{-6}	3×10^{-4}	4×10^{-8}
	(insol.)	8×10^{-3}	10^{-7}	3×10^{-4}	4×10^{-9}
Silicon 31	(sol.)	0.03	6×10^{-6}	9×10^{-4}	2×10^{-7}
	(insol.)	6×10^{-3}	10^{-6}	2×10^{-4}	3×10^{-8}
Silver 105	(sol.)	3×10^{-3}	6×10^{-7}	10^{-4}	2×10^{-8}
	(insol.)	3×10^{-3}	8×10^{-8}	10^{-4}	3×10^{-9}
Silver 110m	(sol.)	9×10^{-4}	2×10^{-7}	3×10^{-5}	7×10^{-9}
	(insol.)	9×10^{-4}	10^{-8}	3×10^{-5}	3×10^{-10}
Silver 111	(sol.)	10^{-3}	3×10^{-7}	4×10^{-5}	10^{-8}
	(insol.)	10^{-3}	2×10^{-7}	4×10^{-5}	8×10^{-9}
Sodium 22	(sol.)	10^{-3}	2×10^{-7}	4×10^{-5}	6×10^{-9}
	(insol.)	9×10^{-4}	9×10^{-9}	3×10^{-5}	3×10^{-10}
Sodium 24	(sol.)	6×10^{-3}	10^{-6}	2×10^{-4}	4×10^{-8}
	(insol.)	8×10^{-4}	10^{-7}	3×10^{-5}	5×10^{-9}
Strontium 85m	(sol.)	0.2	4×10^{-5}	0.007	10^{-6}
	(insol.)	0.2	3×10^{-5}	0.007	10^{-6}
Strontium 85	(sol.)	3×10^{-3}	2×10^{-7}	10^{-4}	8×10^{-9}
	(insol.)	5×10^{-3}	10^{-7}	2×10^{-4}	4×10^{-9}
Strontium 89	(sol.)	3×10^{-4}	3×10^{-8}	3×10^{-6}	3×10^{-10}
	(insol.)	8×10^{-4}	4×10^{-8}	3×10^{-5}	10^{-9}
Strontium 90	(sol.)	1×10^{-5}	1×10^{-9}	4×10^{-7}	4×10^{-11}
	(insol.)	10^{-3}	5×10^{-9}	4×10^{-5}	2×10^{-10}
Strontium 91	(sol.)	2×10^{-3}	4×10^{-7}	7×10^{-5}	2×10^{-8}
	(insol.)	10^{-3}	3×10^{-7}	5×10^{-5}	9×10^{-9}
Strontium 92	(sol.)	2×10^{-3}	4×10^{-7}	7×10^{-5}	2×10^{-8}
	(insol.)	2×10^{-3}	3×10^{-7}	6×10^{-5}	10^{-8}
Sulfur 35	(sol.)	2×10^{-3}	3×10^{-7}	6×10^{-5}	9×10^{-9}
	(insol.)	8×10^{-3}	3×10^{-7}	3×10^{-4}	9×10^{-9}

(continued)

Radionuclide		Occupational 40-hr. Week		Non-Occupational	
		Water uc/ml	Air uc/ml	Water uc/ml	Air uc/ml
Column		A	B	C	D
Tantalum 182	(sol.)	10^{-3}	4×10^{-8}	4×10^{-5}	10^{-9}
	(insol.)	10^{-3}	2×10^{-8}	4×10^{-5}	7×10^{-10}
Technetium 96m	(sol.)	0.4	8×10^{-5}	0.01	3×10^{-6}
	(insol.)	0.3	3×10^{-5}	0.01	10^{-6}
Technetium 96	(sol.)	3×10^{-3}	6×10^{-7}	10^{-4}	2×10^{-8}
	(insol.)	10^{-3}	2×10^{-7}	5×10^{-5}	8×10^{-9}
Technetium 97m	(sol.)	0.01	2×10^{-6}	4×10^{-4}	8×10^{-9}
	(insol.)	5×10^{-3}	2×10^{-7}	2×10^{-4}	5×10^{-9}
Technetium 97	(sol.)	0.05	10^{-5}	0.002	4×10^{-7}
	(insol.)	0.02	3×10^{-7}	8×10^{-4}	10^{-8}
Technetium 99m	(sol.)	0.2	4×10^{-5}	0.006	10^{-6}
	(insol.)	0.08	10^{-5}	0.003	5×10^{-7}
Technetium 99	(sol.)	0.01	2×10^{-6}	3×10^{-4}	7×10^{-8}
	(insol.)	5×10^{-3}	6×10^{-8}	2×10^{-4}	2×10^{-9}
Tellurium 125	(sol.)	5×10^{-3}	4×10^{-7}	2×10^{-4}	10^{-8}
	(insol.)	3×10^{-3}	10^{-7}	10^{-4}	4×10^{-9}
Tellurium 127m	(sol.)	2×10^{-3}	10^{-7}	6×10^{-5}	5×10^{-9}
	(insol.)	2×10^{-3}	4×10^{-8}	5×10^{-5}	10^{-9}
Tellurium 127	(sol.)	8×10^{-3}	2×10^{-6}	3×10^{-4}	6×10^{-8}
	(insol.)	5×10^{-3}	9×10^{-7}	2×10^{-4}	3×10^{-8}
Tellurium 129m	(sol.)	10^{-3}	8×10^{-8}	3×10^{-5}	3×10^{-9}
	(insol.)	6×10^{-4}	3×10^{-8}	2×10^{-5}	10^{-9}
Tellurium 129	(sol.)	0.02	5×10^{-6}	8×10^{-4}	2×10^{-7}
	(insol.)	0.02	4×10^{-6}	8×10^{-4}	10^{-7}
Tellurium 131m	(sol.)	2×10^{-3}	4×10^{-7}	6×10^{-5}	10^{-8}
	(insol.)	10^{-3}	2×10^{-7}	4×10^{-5}	6×10^{-9}
Tellurium 132	(sol.)	9×10^{-4}	2×10^{-7}	3×10^{-5}	7×10^{-9}
	(insol.)	6×10^{-4}	10^{-7}	2×10^{-5}	4×10^{-9}
Terbium 160	(sol.)	10^{-3}	10^{-7}	4×10^{-5}	3×10^{-9}
	(insol.)	10^{-3}	3×10^{-8}	4×10^{-5}	10^{-9}
Thallium 200	(sol.)	0.01	3×10^{-6}	4×10^{-4}	9×10^{-8}
	(insol.)	7×10^{-3}	10^{-6}	2×10^{-4}	4×10^{-8}
Thallium 201	(sol.)	9×10^{-3}	2×10^{-6}	3×10^{-4}	7×10^{-8}
	(insol.)	5×10^{-3}	9×10^{-7}	2×10^{-4}	3×10^{-8}
Thallium 202	(sol.)	4×10^{-3}	8×10^{-7}	10^{-4}	3×10^{-8}
	(insol.)	2×10^{-3}	2×10^{-7}	7×10^{-5}	8×10^{-9}
Thallium 204	(sol.)	3×10^{-3}	6×10^{-7}	10^{-4}	2×10^{-8}
	(insol.)	2×10^{-3}	3×10^{-8}	6×10^{-5}	9×10^{-10}

(continued)

Radionuclide		Occupational 40-hr. Week		Non-Occupational	
		Water uc/ml	Air uc/ml	Water uc/ml	Air uc/ml
Column		A	B	C	D
Thorium 227	(sol.)	5×10^{-4}	3×10^{-10}	2×10^{-5}	10^{-11}
	(insol.)	5×10^{-4}	2×10^{-10}	2×10^{-5}	6×10^{-12}
Thorium 228	(sol.)	2×10^{-4}	9×10^{-12}	7×10^{-6}	3×10^{-13}
	(insol.)	4×10^{-4}	6×10^{-12}	10^{-5}	2×10^{-13}
Thorium 230	(sol.)	5×10^{-5}	2×10^{-12}	2×10^{-6}	8×10^{-14}
	(insol.)	9×10^{-4}	10^{-11}	3×10^{-5}	3×10^{-13}
Thorium 231	(sol.)	7×10^{-3}	10^{-6}	2×10^{-4}	5×10^{-8}
	(insol.)	7×10^{-3}	10^{-6}	2×10^{-4}	4×10^{-8}
Thorium 232	(sol.)	5×10^{-5}	3×10^{-11}	2×10^{-6}	10^{-12}
	(insol.)	10^{-3}	3×10^{-11}	4×10^{-5}	10^{-12}
Thorium 234	(sol.)	5×10^{-4}	6×10^{-8}	2×10^{-5}	2×10^{-9}
	(insol.)	5×10^{-4}	3×10^{-8}	2×10^{-5}	10^{-9}
Thorium Nat.	(sol.)	3×10^{-5}	3×10^{-11}	10^{-6}	10^{-12}
	(insol.)	3×10^{-4}	3×10^{-11}	10^{-5}	10^{-12}
Thulium 170	(sol.)	10^{-3}	4×10^{-8}	5×10^{-5}	10^{-9}
	(insol.)	10^{-3}	3×10^{-8}	5×10^{-5}	10^{-9}
Thulium 171	(sol.)	0.01	10^{-7}	5×10^{-4}	4×10^{-9}
	(insol.)	0.01	2×10^{-7}	5×10^{-4}	8×10^{-9}
Tin 113	(sol.)	2×10^{-3}	4×10^{-7}	9×10^{-5}	10^{-8}
	(insol.)	2×10^{-3}	5×10^{-8}	8×10^{-5}	2×10^{-9}
Tin 125	(sol.)	5×10^{-4}	10^{-7}	2×10^{-5}	4×10^{-9}
	(insol.)	5×10^{-4}	8×10^{-8}	2×10^{-5}	3×10^{-9}
Tungsten 181	(sol.)	0.01	2×10^{-6}	4×10^{-4}	8×10^{-8}
	(insol.)	0.01	10^{-7}	3×10^{-4}	4×10^{-9}
Tungsten 185	(sol.)	4×10^{-3}	8×10^{-7}	10^{-4}	3×10^{-8}
	(insol.)	3×10^{-3}	10^{-7}	10^{-4}	4×10^{-9}
Tungsten 187	(sol.)	2×10^{-3}	4×10^{-7}	7×10^{-5}	2×10^{-8}
	(insol.)	2×10^{-3}	3×10^{-7}	6×10^{-5}	10^{-8}
Uranium 230	(sol.)	7×10^{-5}	3×10^{-10}	2×10^{-6}	10^{-11}
	(insol.)	10^{-4}	10^{-10}	5×10^{-6}	4×10^{-12}
Uranium 232	(sol.)	2×10^{-5}	10^{-10}	8×10^{-7}	3×10^{-12}
	(insol.)	8×10^{-4}	3×10^{-11}	3×10^{-5}	9×10^{-13}
Uranium 233	(sol.)	10^{-4}	5×10^{-10}	4×10^{-6}	2×10^{-11}
	(insol.)	9×10^{-4}	10^{-10}	3×10^{-5}	4×10^{-12}
Uranium 234	(sol.)	10^{-4}	6×10^{-10}	4×10^{-6}	2×10^{-11}
	(insol.)	9×10^{-4}	10^{-10}	3×10^{-5}	4×10^{-12}
Uranium 235	(sol.)	10^{-4}	5×10^{-10}	4×10^{-6}	2×10^{-11}
	(insol.)	8×10^{-4}	10^{-10}	3×10^{-5}	4×10^{-12}

(continued)

Radionuclide		Occupational 40-hr. Week		Non-Occupational	
		Water uc/ml	Air uc/ml	Water uc/ml	Air uc/ml
Column		**A**	**B**	**C**	**D**
Uranium 236	(sol.)	10^{-4}	6×10^{-10}	5×10^{-6}	2×10^{-11}
	(insol.)	10^{-3}	10^{-10}	3×10^{-5}	4×10^{-12}
Uranium 238	(sol.)	2×10^{-5}	7×10^{-11}	6×10^{-7}	3×10^{-12}
	(insol.)	10^{-3}	10^{-10}	4×10^{-5}	5×10^{-12}
Uranium 240	(sol.)	1×10^{-3}	2×10^{-7}	3×10^{-5}	8×10^{-9}
& Neptunium 240	(insol.)	1×10^{-3}	2×10^{-7}	3×10^{-5}	6×10^{-9}
Uranium-Nat.	(sol.)	2×10^{-5}	7×10^{-11}	6×10^{-7}	3×10^{-12}
	(insol.)	5×10^{-4}	6×10^{-11}	2×10^{-5}	2×10^{-12}
Vanadium 48	(sol.)	9×10^{-4}	2×10^{-7}	3×10^{-5}	6×10^{-9}
	(insol.)	8×10^{-4}	6×10^{-8}	3×10^{-5}	2×10^{-9}
Xenon 131m	(imm.)	....	2×10^{-5}	....	4×10^{-7}
Xenon 133	(imm.)	....	10^{-5}	....	3×10^{-7}
Xenon 133m	(imm.)	....	1×10^{-5}	....	3×10^{-7}
Xenon 135	(imm.)	....	4×10^{-6}	....	10^{-7}
Ytterbium 175	(sol.)	3×10^{-3}	7×10^{-7}	10^{-4}	2×10^{-8}
	(insol.)	3×10^{-3}	6×10^{-7}	10^{-4}	2×10^{-8}
Yttrium 90	(sol.)	6×10^{-4}	10^{-7}	2×10^{-5}	4×10^{-9}
	(insol.)	6×10^{-4}	10^{-7}	2×10^{-5}	3×10^{-9}
Yttrium 91m	(sol.)	0.1	2×10^{-5}	0.003	8×10^{-7}
	(insol.)	0.1	2×10^{-5}	0.003	6×10^{-7}
Yttrium 91	(sol.)	8×10^{-4}	4×10^{-8}	3×10^{-5}	10^{-9}
	(insol.)	8×10^{-4}	3×10^{-8}	3×10^{-5}	10^{-9}
Yttrium 92	(sol.)	2×10^{-3}	4×10^{-7}	6×10^{-5}	10^{-8}
	(insol.)	2×10^{-3}	3×10^{-7}	6×10^{-5}	10^{-8}
Yttrium 93	(sol.)	8×10^{-4}	2×10^{-7}	3×10^{-5}	6×10^{-9}
	(insol.)	8×10^{-4}	10^{-7}	3×10^{-5}	5×10^{-9}
Zinc 65	(sol.)	3×10^{-3}	10^{-7}	10^{-4}	4×10^{-9}
	(insol.)	5×10^{-3}	6×10^{-8}	2×10^{-4}	2×10^{-9}
Zinc 69m	(sol.)	2×10^{-3}	4×10^{-7}	7×10^{-5}	10^{-8}
	(insol.)	2×10^{-3}	3×10^{-7}	6×10^{-5}	10^{-8}
Zinc 69	(sol.)	0.05	7×10^{-6}	0.002	2×10^{-7}
	(insol.)	0.05	9×10^{-6}	0.002	3×10^{-7}
Zirconium 93	(sol.)	0.02	10^{-7}	8×10^{-4}	4×10^{-9}
	(insol.)	0.02	3×10^{-7}	8×10^{-4}	10^{-8}
Zirconium 95	(sol.)	2×10^{-3}	10^{-7}	6×10^{-5}	4×10^{-9}
	(insol.)	2×10^{-3}	3×10^{-8}	6×10^{-5}	10^{-9}
Zirconium 97	(sol.)	5×10^{-4}	10^{-7}	2×10^{-5}	4×10^{-9}
	(insol.)	5×10^{-4}	9×10^{-8}	2×10^{-5}	3×10^{-9}
Unidentified Radionuclide(s)		3×10^{-7}	1×10^{-12}	10^{-8}	4×10^{-14}

Abbreviations — sol. = soluble   imm. = immersion
insol. = insoluble   m = metastable

NOTE: In any case where there is a mixture in air or water of more than one radio-nuclide, the limiting values for purposes of this section shall be determined as follows:

If the identity and concentration of each radionuclide in the mixture are known, the limiting values shall be derived as follows: Determine for each radionuclide in the mixture, the ratio between the quantity present in the mixture and the limit otherwise established in this Section for the specific radionuclide when not in a mixture. The sum of such ratios for all the radionuclides in the mixture may not exceed "1" (i.e., "unity").

EXAMPLE: If radionuclides A, B, and C are present in concentrations C_a, C_b and C_c, and if the applicable MPC's are MPC_a and MPC_b and MPC_c, respectively, then the concentrations shall be limited so that the following relationship exists:

$$\frac{C_a}{\text{MPC}_a} + \frac{C_b}{\text{MPC}_b} + \frac{C_c}{\text{MPC}_c} \leq 1$$

GLOSSARY

Accelerator   a device for imparting very high velocity to charged particles such as electrons or protons. These fast particles can penetrate matter and are known as radiation. Fast particles of this type are used in research or to study the structure of the atom itself.

Alpha Particle (Alpha Ray, Alpha Radiation)   a small electrically charged particle of very high velocity thrown off by many radioactive materials, including uranium and radium. It is identical with the nucleus of a helium atom and is made up of two neutrons and two protons. Its electric charge is positive and twice as great as that of an electron.

Atom   the tiny "building block" of nature. All materials are made of atoms. The elements, such as iron, lead, and sulfur, differ from each other because they contain different atoms. Atoms are unbelievably small. No one has ever seen one. There are six sextillion (6 followed by 21 zeros) atoms in an ordinary drop of water. The word "atom" comes from the Greek word meaning indivisible. Now we know it can be split and consists of an inner core (nucleus) surrounded by electrons which rotate around the nucleus like the planets around the sun.

Atomic Energy   energy released in nuclear reactions. Of particular interest is the energy released when a neutron splits an atom's nucleus into smaller pieces (fission) or when two nuclei are joined together under millions of degrees of heat (fusion). "Atomic energy" is really a popular misnomer. It is more correctly called *nuclear energy*.

Atomic Number   the number of protons (positively charged particles) found in the nucleus of an atom. All elements have different atomic numbers. The atomic number of hydrogen is 1, that of oxygen 8, iron 26, lead 82, uranium 92.

Atomic Theory   since the time of the ancient Greeks man has held the theory that all matter is composed of tiny, invisible particles called atoms. It remained for the chemists and physicists of the 19th and 20th centuries to verify the existence of the atom and the validity of the atomic theory.

Atomic Weight   the atomic weight is approximately the sum of the number of protons and neutrons found in the nucleus of an atom. The atomic weight of oxygen, for example, is approximately 16 (actually it is 16.0044)—it contains 8 neutrons plus 8 protons. Aluminum is 27—it contains 14 neutrons and 13 protons.

Atom Smasher   a machine (an accelerator) that speeds up atomic and subatomic particles so that they can be used as projectiles to literally blast apart the nuclei of other atoms.

Autoradiography   self-portraits of radioactive sources made by placing the radioactive material next to photographic film. The radiations fog the film leaving an image of the source. It was such self-portraits that led to the discovery of radioactivity.

Background   background radiation is always detected by a counter. It is caused by radiation coming from sources other than the radioactive material to be measured. This "background" is primarily due to cosmic rays which constantly bombard the earth from outer space.

Beta Particle (Beta Radiation)   a small electrically charged particle thrown off by many radioactive materials. It is identical with the electron and possesses the smallest negative electric charge found in nature. Beta particles emerge from radioactive material at high speeds, sometimes close to the speed of light.

Betatron   a large doughnut-shaped accelerator in which electrons (beta particles) are whirled through a changing magnetic field gaining speed with each trip and emerging with high energies. Energies of the order of 100 million electron volts have been achieved. The betatron produces artificial beta radiation.

Binding Energy   the energy which holds the neutrons and protons of an atomic nucleus together.

Bombardment   shooting neutrons, alpha particles, and other high energy particles at atomic nuclei usually in an attempt to split the nucleus or to form a new element.

Breeder   a reactor which is producing more atomic fuel than it is consuming. A nonfissionable isotope, bombarded by neutrons, is transformed into a fissionable material, such as plutonium, which can be used as fuel. Scientists are working toward the day when all the material burned in reactors will be replaced through this process.

Cerenkov Radiation   an eerie blue glow given off by electrons traveling in a transparent material such as water. It is this radiation which is visible during the operation of some nuclear reactors.

Chain Reaction   when a fissionable nucleus is split by a neutron it releases energy and one or more neutrons. These neutrons split other fissionable nuclei releasing more energy and more neutrons making the reaction self-sustaining.

Charge   the fuel (fissionable material) placed in a reactor to produce a chain reaction.

Cloud Chamber   a glass-domes chamber filled with moist vapor. When certain types of atomic particles pass through the chamber they leave a cloud-like track much like the vapor trail of a jet plane. This permits scientists to "see" these particles and study their motion.

Cobalt-60   a radioactive isotope of the element cobalt. Cobalt-60 is an important source of gamma radiation and is used widely in research.

Coffin   a thick-walled container (usually lead) used for transporting radioactive materials.

Control Rod   a rod used to control the power of a nuclear reactor. The reactor functions through the splitting of nuclear fuel by neutrons. The control rod absorbs neutrons which would normally split atoms of the fuel. Pushing the rod in reduces the release of atomic power. Pulling out the rod increases it.

Converter   a reactor which uses one kind of fuel and produces another. For example, a converter charged with uranium isotopes might consume uranium-235 and product plutonium from uranium-238.

Core   the heart of a nuclear reactor where the nuclei of the fuel fission (split) and release energy. The core is usually surrounded by a reflecting material which bounces stray neutrons back to the fuel.

Cosmotron   a huge accelerator, one of the atomic "guns" located at Brookhaven National Laboratory. It speeds up particles to the billion electron volt range. The Brookhaven machine has a magnet weighing 2,200 tons.

Counter   a device for counting nuclear disintegrations to measure radioactivity. The signal which announces a disintegration is called a *count*.

Critical Mass   the amount of nuclear fuel necessary to sustain a chain reaction. If too little fuel is present too many neutrons will stray and the reaction will die out.

Curie   a measure of the rate at which a radioactive material throws off particles. The radioactivity of one gram of radium is a curie. It is named for Pierre and Marie Curie, pioneers in radioactivity and discoverers of the elements radium, radon, and polonium.

Cutie-Pie   a portable instrument equipped with a direct reading meter used to determine the level of radiation in an area.

Cyclotron   a particle accelerator. In this atomic "merry-go-round" atomic particles are whirled around in a spiral between the ends of a huge magnet gaining speed with each rotation in preparation for their assault on the target material.

Decay   when a radioactive atom disintegrates it is said to decay. What remains is a different element. An atom of polonium decays to form lead, ejecting an alpha particle in the process.

Deuterium   heavy hydrogen. The nucleus of heavy hydrogen is a deuteron. It is called heavy hydrogen because it weighs twice as much as ordinary hydrogen.

Dosimeter (Dose Meter)   an instrument used to determine the radiation dose a person has received.

Electron   a minute atomic particle possessing the smallest amount of negative electric charge found in nature. In an atom the electrons rotate around a small nucleus. The weight of an electron is so infinitesimal that it would take 500 octillions (500 followed by 27 zeros) of them to make a pound. It is only about a two-thousandth of the mass of a proton or neutron.

Element   a basic substance consisting of a family of naturally occurring isotopes.

For example, hydrogen, lead, and oxygen are elements. All atoms of an element contain a definite number of protons and thus have the same atomic number.

Film Badge   a piece of masked photographic film worn like a badge by nuclear workers. It is darkened by nuclear radiation, and radiation exposure can be checked by inspecting the film.

Fission   the splitting of an atomic nucleus into two parts accompanied by the release of a large amount of radioactivity and heat. Fission reactions occur only with heavy elements such as uranium and plutonium.

Fusion   the joining of atomic nuclei to form a heavier nucleus, accomplished under conditions of extreme heat (millions of degrees). If two nuclei of light atoms fuse, the fusion is accompanied by the release of a great deal of energy. The energy of the sun is believed to be derived from the fusion of hydrogen atoms to form helium.

Gamma Rays (Gamma Radiation)   the most penetrating of all radiations. Gamma rays are very high energy X-rays.

Geiger Counter   a gas-filled electrical device which detects the presence of radioactivity by counting the formation of ions.

Half-Life   a means of classifying the rate of decay of radioisotopes according to the time it takes them to lose half their strength (intensity). Half-lives range from fractions of seconds to billions of years. Cobalt-60, for example, has a half-life of 5.3 years. A radioactive material loses half its strength when its age is equal to its half-life.

Heavy Hydrogen   same as deuterium.

Heavy Water   water which contains heavy hydrogen (deuterium) instead of ordinary hydrogen. It is widely used in reactors to slow down neutrons.

Ion   usually an atom which has lost one or more of its electrons and is left with a positive electrical charge. There are also negative ions, which have gained an extra electron.

Ionization Chamber   a device roughly similar to a Geiger counter and used to measure radioactivity.

Isotope   two nuclei of the same element which have the same charge but different masses are called isotopes. They contain the same number of protons but a different number of neutrons. Uranium-238 contains 92 protons and 146 neutrons while the isotope U-235 contains 92 protons and 143 neutrons. Thus, the atomic weight (atomic mass) of U-238 is three higher than of U-235.

Kilocurie   1,000 curies. A unit of radioactivity.

Master Slave Manipulators   mechanical hands used to handle hot materials. They are remotely controlled from behind a protective shield.

Meson   a particle which weighs more than the electron but generally less than the proton. Mesons can be produced artificially. They are also produced by cosmic radiation (natural radiation coming from outer space). Mesons are not stable— they disintegrate in a fraction of a second.

Milliroentgen   one one-thousandth of a roentgen. A unit of radioactive dose.

Moderator   a material used to slow neutrons in a reactor. These slow neutrons are particularly effective in causing fission. Neutrons are slowed down when they collide with atoms of light elements such as hydrogen and carbon, two common moderators.

Molecule   the smallest unit of a compound. A water molecule consists of two hydrogen atoms combined with one oxygen atom. Hence, the well-known formula, H_2O.

Monitor   a radiation detector used to determine whether an area is safe for workers. A cutie-pie is a portable monitor.

Neutron   one of the three basic atomic particles. The neutron weighs about the same as the proton and, as its name implies, has no electric charge. Neutrons make effective atomic projectiles.

Nuclear Bombardment   the shooting of atomic projectiles at nuclei usually in an attempt to split the atom or to form a new element.

Nuclear Energy   the energy released in a nuclear reaction, such as fission or fusion. Nuclear energy is popularly, though mistakenly, called atomic energy.

Nuclear Reaction   result of the bombardment of a nucleus with atomic or subatomic particles or very high energy radiation. Possible reactions are emission of other particles or the splitting of the nucleus (fission). The decay of a radioactive material is also a nuclear reaction.

Nucleus   the inner core of the atom. It consists of neutrons and protons tightly locked together.

Photoelectric Effect   occurs when an electron is knocked out of an atom by a light ray or gamma ray. This effect is used in an electric eye. Light falls on a sensitive surface knocking out electrons which can then be detected.

Photon   a bundle (quantum) of radiation. Constitutes, for example, X-rays and light. In certain processes gamma rays behave like photons.

Pile   a nuclear reactor. Called a pile because the earliest reactors were "piles" of graphite blocks and uranium slugs.

Pitchblende   an ore containing both uranium and radium. The Curies had to purify tons of pitchblende to obtain a barely visible speck of radium.

Plutonium   a heavy element which undergoes fission under the impact of neutrons. It is a useful fuel in nuclear reactors. Plutonium does not occur in nature but can be produced and "burned" in reactors.

Positron   a particle which has the same weight and charge as an electron but is electrically positive rather than negative. The positron's existence was predicted in theory years before it was actually detected. It is not stable in matter.

Proton   one of the basic particles of the atomic nucleus (the other is the neutron). Its charge is as large as that of the electron, but positive.

Radiation (Radioactivity)   the emission of very fast atomic particles or rays by nuclei. Some elements are naturally radioactive while others become radioactive after bombardment with neutrons or other particles. The three major forms of radiation are alpha, beta, and gamma, named for the first three letters of the Greek alphabet.

Radioisotope   a radioactive isotope of an element. A radioisotope can be produced by placing material in a nuclear reactor and bombarding it with neutrons. Radio-isotopes are being used today as tracers in many areas of science and industry and are at present the most important peacetime contribution of atomic energy.

Radium   one of the earliest known naturally radioactive elements. It is far more radioactive than uranium and is found in the same ores.

Reactor   an atomic "furnace." In a reactor, nuclei of the fuel undergo fission under the influence of neutrons. The fission produces new neutrons, and hence, a chain reaction. This releases large amounts of energy. This energy is removed as heat which may be used to make steam for use in generation of electricity.

Roentgen   a unit of radioactive dose, or exposure. The Atomic Energy Commission has established a conservative limit of exposure for the protection of atomic workers.

Shield   a wall which protects workers from harmful radiations released by radioactive materials.

Slug   a "fuel element" for a nuclear reactor, a piece of fissionable material. The slugs in large reactors consist of uranium metal coated with aluminum to prevent corrosion.

Source   any substance which emits radiation. Usually refers to a piece of radioactive material conveniently packaged for scientific or industrial use.

Thermonuclear Reaction   a fusion reaction, that is, a reaction in which two light nuclei combine to form a heavier atom, releasing a large amount of energy. This is believed to be the sun's source of energy. It is called thermonuclear because it occurs only at a very high temperature.

Thorium   a heavy element. When bombarded with neutrons, thorium changes into uranium, becoming fissionable and thus a source of atomic energy.

Tracer   a radioisotope which is mixed with a stable material. The radioisotope enables scientists to trace the material as it undergoes chemical and physical changes. Tracers are being used widely in science, industry, and agriculture today. When radioactive phosphorous, for example, is mixed with a chemical fertilizer the radioactive substance can be traced through the plant as it grows.

Tritium   often called hydrogen three. Extra heavy hydrogen whose nucleus contains two neutrons and one proton. It is three times as heavy as ordinary hydrogen and is radioactive.

Unstable   all radioactive elements are unstable since they emit particles and decay to form other elements.

Uranium   a heavy metal. The two principal isotopes of natural uranium are U-235 and U-238. U-235 has the only readily fissionable nucleus which occurs in appreciable quantities in nature, hence, its importance as nuclear fuel. Only 1 part in 140 of natural uranium is U-235.

Van de Graaff Accelerator   an electrostatic generator—a particle accelerator. To obtain the voltage, static electricity is picked up at one end of the machine by a rubber belt and carried to the other end where it is stored.

X-Ray   highly penetrating radiation similar to gamma rays. Unlike gamma rays, X-rays do not come from the nucleus of the atom but from the surrounding electrons. They are produced by electron bombardment. When these rays pass through an object they give a shadow picture of the denser portions.

INDEX